Hydroblasting

Trainee Guide
Second Edition

PEARSON

Boston   Columbus   Indianapolis   New York   San Francisco   Upper Saddle River
Amsterdam   Cape Town   Dubai   London   Madrid   Milan   Munich   Paris   Montreal   Toronto
Delhi   Mexico City   São Paulo   Sydney   Hong Kong   Seoul   Singapore   Taipei   Tokyo

National Center for Construction Education and Research

President: Don Whyte
Director of Product Development: Daniele Stacey
Hydroblasting Project Manager: Matt Tischler
Production Manager: Tim Davis
Quality Assurance Coordinator: Debie Ness

Desktop Publishing Coordinator: James McKay
Production Specialist: Megan Casey
Editor: Chris Wilson

Writing and development services provided by Integra Software Services Pvt. Ltd

Project Manager: Allison Campbell
Writer: Paul Klemt

Pearson Education, Inc.

Editorial Director: Vernon R. Anthony
Executive Editor: Alli Gentile
Senior Product Manager: Lori Cowen
Operations Supervisor: Deidra M. Skahill
Art Director: Jayne Conte

Director of Marketing: David Gesell
Executive Marketing Manager: Derril Trakalo
Marketing Manager: Brian Hoehl
Marketing Coordinator: Crystal Gonzalez

Composition: NCCER
Printer/Binder: Document Technology Resources, Fredericksburg, VA
Cover Printer: Document Technology Resources, Fredericksburg, VA
Text Fonts: Palatino and Univers

Perfect bound   ISBN-13: 978-0-13-294870-8
ISBN-10:   0-13-294870-2

Preface

To the Trainee

Hydroblasters require a specific set of skills, and must have knowledge of water pressure, cleaning of various materials, and jobsite safety. Once properly trained, your skills as a hydroblaster will afford you many prospects for success in industry.

Jets of high-pressure water, often between 10,000 and 20,000 psi, are used to clean tanks, boilers, remove paint, and perform numerous maintenance functions necessary for high-quality operations in industrial applications. Safety is essential in this trade, and you will wear special protective clothing, such as gloves, hard hats, respirators, goggles, and other devices. You will work closely with others, develop strong team relationships, and learn to communicate effectively. You will also develop problem-solving skills to evaluate the difficulty of a job and determine the proper procedure to address any issues.

This second edition of NCCER's 20 hour *Hydroblasting* curriculum is now in full color and includes updates to safety standards, hydroblasting tools, and new waterjet technologies. Cold and warm weather safety coverage has been expanded, automatic hydroblasting equipment is featured, and ultra-high pressure applications are explained.

As safety and best practices have become a priority over the past several years, hydroblasting has become recognized as a skilled craft and the need for hydroblasting services has increased. This curriculum will provide the specialized training necessary to perform a variety of tasks based on today's industry needs. Prospects are bright in this field; with a positive attitude and an excellent work ethic, you can steadily advance in industrial cleaning and find gainful employment in a variety of settings and locations.

We wish you success as you progress through this training program. Should you have any comments on how NCCER might improve upon this textbook, please complete the User Update form located at the back of this module and send it to us. We will always consider and respond to input from our customers.

We invite you to visit the NCCER website at **www.nccer.org** for information on the latest product releases and training, as well as online versions of the *Cornerstone* newsletter and Pearson's NCCER product catalog.

Your feedback is welcome. You may email your comments to **curriculum@nccer.org** or send general comments and inquiries to **info@nccer.org**.

NCCER Standardized Curricula

NCCER is a not-for-profit 501(c)(3) education foundation established in 1996 by the world's largest and most progressive construction companies and national construction associations. It was founded to address the severe workforce shortage facing the industry and to develop a standardized training process and curricula. Today, NCCER is supported by hundreds of leading construction and maintenance companies, manufacturers, and national associations. The NCCER Standardized Curricula was developed by NCCER in partnership with Pearson, the world's largest educational publisher.

Some features of the NCCER Standardized Curricula are as follows:

- An industry-proven record of success
- Curricula developed by the industry for the industry
- National standardization providing portability of learned job skills and educational credits
- Compliance with the Office of Apprenticeship requirements for related classroom training (*CFR 29:29*)
- Well-illustrated, up-to-date, and practical information

NCCER also maintains a National Registry that provides transcripts, certificates, and wallet cards to individuals who have successfully completed modules of the NCCER Standardized Curricula. *Training programs must be delivered by an NCCER Accredited Training Sponsor in order to receive these credentials.*

Special Features

In an effort to provide a comprehensive, user-friendly training resource, we have incorporated many different features for your use. Whether you are a visual or hands-on learner, this book will provide you with the proper tools to get started in the field of hydroblasting.

Introduction

This page is found at the beginning of each module and lists the Objectives, Performance Tasks, and Trade Terms for that module. The Objectives list the skills and knowledge you will need in order to complete the module successfully. The Performance Tasks give you the opportunity to apply your knowledge to the real world duties that hydroblasting technicians perform. The list of Trade Terms identifies important terms you will need to know by the end of the module.

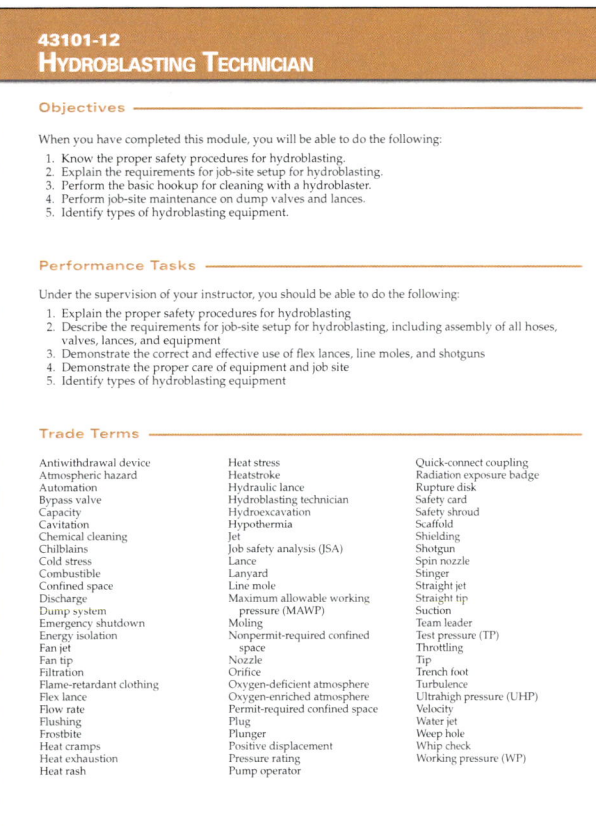

Color Illustrations and Photographs

Full-color illustrations and photographs are used throughout each module to provide vivid detail. These figures highlight important concepts from the text and provide clarity for complex instructions. Each figure reference is denoted in the text in *italic type* for easy reference.

Figure 44 Automated equipment.

Notes, Cautions, and Warnings

Safety features are set off from the main text in highlighted boxes and are organized into three categories based on the potential danger of the issue being addressed. Notes simply provide additional information on the topic area. Cautions alert you of a danger that does not present potential injury but may cause damage to equipment. Warnings stress a potentially dangerous situation that may cause injury to you or a co-worker.

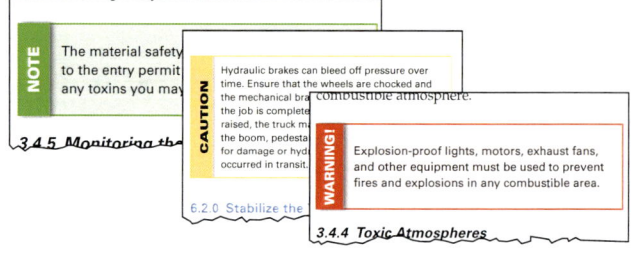

Case History

Case History features emphasize the importance of safety by citing examples of the costly (and often devastating) consequences of ignoring best practices or OSHA regulations.

Case History

Added Oxygen Causes Lost Life

A blaster went into a confined space to waterblast a tanker. The tanker had VCM (vinyl chloride monomer) that mixed with water and turned into carbon monoxide and displaced the oxygen. He died and the whole crew that followed him in died as well.

The Bottom Line: Make sure you know what you're going into clean, review the SDS, ensure the PPE is correct, and verify the rescue plan. Do not rescue the worker yourself. This accident could have been avoided. It happened because of poor communication between workers and unsafe work practices.

Step-by-Step Instructions

Step-by-step instructions are used throughout to guide you through technical procedures and tasks from start to finish. These steps show you not only how to perform a task but also how to do it safely and efficiently.

Follow these steps to hook up the lance:

Step 1 Connect the pressure hose to the pump (*Figure 59*). Push the coupling closed and give the hose a tug to be sure it is firmly closed. When pressure hoses are connected with quick-connect couplings, they must also be provided with anti-whip cables. These cables are placed so as to prevent the hoses from whipping around in case of accidental release. There is a great deal of recoil from water being discharged at pressure from the hose. The whip checks hold the parts of the hose together, and prevent the coupling and hose from whipping around and possibly striking and injuring personnel.

Step 2 Make sure that the pressure control (*Fig-*

Trade Terms

Each module presents a list of Trade Terms that are discussed within the text and defined in the Glossary at the end of the module. These terms are denoted in the text with bold, blue type upon their first occurrence.

---

to a cool, shaded area and sprayed, sponged, or showered with cool water.

Heat exhaustion occurs when the body loses an excessive amount of water and salt, usually due to excessive sweating. Employees at greatest risk for heat exhaustion are older, have high blood pressure, and work in a hot environment. Some symptoms of heat exhaustion are heavy sweating, extreme weakness or fatigue, nausea, moist and/or clammy skin, muscle cramps, and fast and shallow breathing. A co-worker with heat exhaustion should be moved to a cool, shaded or air-conditioned area, drink plenty of water, and take a cool shower, bath, or sponge bath.

Heat cramps usually affect workers who sweat a lot during strenuous activity. Their excessive sweating decreases their bodies' salt and water

---

Review Questions

Review Questions are provided to reinforce the knowledge you have gained. This makes them a useful tool for measuring what you have learned.

Review Questions

1. The range for pressure used in hydroblasting is from _____.
   a. 100 to 2,000 psi
   b. 1,000 to 4,000 psi
   c. 1,000 to 10,000 psi
   d. 1,000 to 48,000 psi

2. Impact is the function of mass and _____.
   a. velocity
   b. pressure
   c. volume
   d. flow rate

3. It is necessary to go to a doctor for a small surface wound from a hydroblasting lance if it *is not* bleeding.
   a. True
   b. False

4. Hydroblasting workers need hearing protection when sound exceeds 85 decibels.
   a. True
   b. False

5. If you are cleaning a tank (from the outside), and you have been told that it is a permit-required confined space, and you drop a nozzle into the tank, you should _____.
   a. climb inside and get it, it'll just take a minute
   b. tell your other operator, then climb in and get it
   c. contact the supervisors at the worksite, and arrange a confined space permit and an entry
   d. try to reach it by sticking your upper body inside

6. Which of the _____ _____ on a water blast _____

7. The type of pump used to produce high-pressure water for hydroblasting is a _____.
   a. positive-displacement pump
   b. centrifugal pump
   c. peristaltic pump
   d. rotary gear pump

8. The dump valve commonly used with a flex lance is a(n) _____.
   a. air relief
   b. hand
   c. foot pedal
   d. knife gate

9. The discharge-side hose should be burst rated at minimum _____.
   a. exactly the expected pressure
   b. one and one-half times the expected pressure
   c. twice the expected pressure
   d. two and one-half times the expected pressure

10. The rigid piece of pipe at the connection of a flex lance to a nozzle is called a(n) _____.
    a. shotgun
    b. stinger
    c. antiwithdrawal device
    d. mouse

11. Fan jet nozzles produce a _____.
    a. 0-degree spray
    b. 15-degree spray
    c. 45-degree spray
    d. 90-degree spray

12. The person in control of the dump system is the one _____.
    a. _____ pump controls

NCCER Standardized Curricula

NCCER's training programs comprise more than 80 construction, maintenance, pipeline, and utility areas and include skills assessments, safety training, and management education.

Boilermaking
Cabinetmaking
Carpentry
Concrete Finishing
Construction Craft Laborer
Construction Technology
Core Curriculum:
    Introductory Craft Skills
Drywall
Electrical
Electronic Systems Technician
Heating, Ventilating, and
    Air Conditioning
Heavy Equipment Operations
Highway/Heavy Construction
Hydroblasting
Industrial Coating and Lining
    Application Specialist
Industrial Maintenance
    Electrical and Instrumentation
    Technician
Industrial Maintenance
    Mechanic
Instrumentation
Insulating
Ironworking
Masonry
Millwright
Mobile Crane Operations
Painting
Painting, Industrial
Pipefitting
Pipelayer
Plumbing
Reinforcing Ironwork
Rigging
Scaffolding
Sheet Metal
Signal Person
Site Layout
Sprinkler Fitting
Tower Crane Operator
Welding

Green/Sustainable Construction

Building Auditor
Fundamentals of Weatherization
Introduction to Weatherization
Sustainable Construction
    Supervisor
Weatherization Crew Chief
Weatherization Technician
Your Role in the Green
    Environment

Energy

Alternative Energy
Introduction to the Power
    Industry
Introduction to Solar
    Photovoltaics
Introduction to Wind Energy
Power Industry Fundamentals
Power Generation Maintenance
    Electrician
Power Generation I&C
    Maintenance Technician
Power Generation Maintenance
    Mechanic
Power Line Worker
Power Line Worker: Distribution
Power Line Worker: Substation
Power Line Worker:
    Transmission
Solar Photovoltaic Systems
    Installer
Wind Turbine Maintenance
    Technician

Pipeline

Control Center Operations,
    Liquid
Corrosion Control
Electrical and Instrumentation
Field Operations, Liquid
Field Operations, Gas
Maintenance
Mechanical

Safety

Field Safety
Safety Orientation
Safety Technology

Management

Fundamentals of Crew
    Leadership
Project Management
Project Supervision

Supplemental Titles

Applied Construction Math
Careers in Construction
Tools for Success

Spanish Translations

Basic Rigging
    (Principios Básicos de
    Maniobras)
Carpentry Fundamentals
    (Introducción a la
    Carpintería, Nivel Uno)
Carpentry Forms
    (Formas para Carpintería,
    Nivel Trés)
Concrete Finishing, Level One
    (Acabado de Concreto,
    Nivel Uno)
Core Curriculum:
    Introductory Craft Skills
    (Currículo Básico:
    Habilidades Introductorias del
    Oficio)
Drywall, Level One
    (Paneles de Yeso, Nivel Uno)
Electrical, Level One
    (Electricidad, Nivel Uno)
Field Safety
    (Seguridad de Campo)
Insulating, Level One
    (Aislamiento, Nivel Uno)
Ironworking, Level One
    (Herrería, Nivel Uno)
Masonry, Level One
    (Albañilería, Nivel Uno)
Pipefitting, Level One
    (Instalación de Tubería
    Industrial, Nivel Uno)
Reinforcing Ironwork, Level One
    (Herreria de Refuerzo,
    Nivel Uno)
Safety Orientation
    (Orientación de Seguridad)
Scaffolding
    (Andamios)
Sprinkler Fitting, Level One
    (Instalación de Rociadores,
    Nivel Uno)

Acknowledgments

This curriculum was revised as a result of the farsightedness and leadership of the following sponsors:

AbClean
Alliance Safety Council
Aquilex HydroChem
Associated Builders and Contractors, Pelican
    Chapter
BASF

Dow
ExxonMobil
FS Solutions Group
Greater Baton Rouge Industry Alliance
TMA Environmental, Inc.
Veolia Environmental Services

This curriculum would not exist were it not for the dedication and unselfish energy of those volunteers who served on the Authoring Team. A sincere thanks is extended to the following:

Sheri Bankston
Evans Barber
Tyler Bargas
Shawn Benoit
Chad Brownell
Mike Champagne
Robert Clouatre
Dwayne Cobb
Connie Fabre
David Hamm

Glenn Miano
Rick Pitman
Trey Rivet
Peter Roblin
Alan Smith
Doris Smith
Don Stewart
David Summers
David Thompson
Gary Toothe

NCCER Partners

American Fire Sprinkler Association
Associated Builders and Contractors, Inc.
Associated General Contractors of America
Association for Career and Technical Education
Association for Skilled and Technical Sciences
Carolinas AGC, Inc.
Carolinas Electrical Contractors Association
Center for the Improvement of Construction
    Management and Processes
Construction Industry Institute
Construction Users Roundtable
Construction Workforce Development Center
Design Build Institute of America
Merit Contractors Association of Canada
Metal Building Manufacturers Association
NACE International
National Association of Minority Contractors
National Association of Women in Construction
National Insulation Association
National Ready Mixed Concrete Association
National Technical Honor Society
National Utility Contractors Association
NAWIC Education Foundation
North American Technician Excellence

Painting & Decorating Contractors of America
Portland Cement Association
Skills USA
Steel Erectors Association of America
The Manufacturers Institute
U.S. Army Corps of Engineers
University of Florida, M. E. Rinker School of
    Building Construction
Women Construction Owners & Executives,
    USA

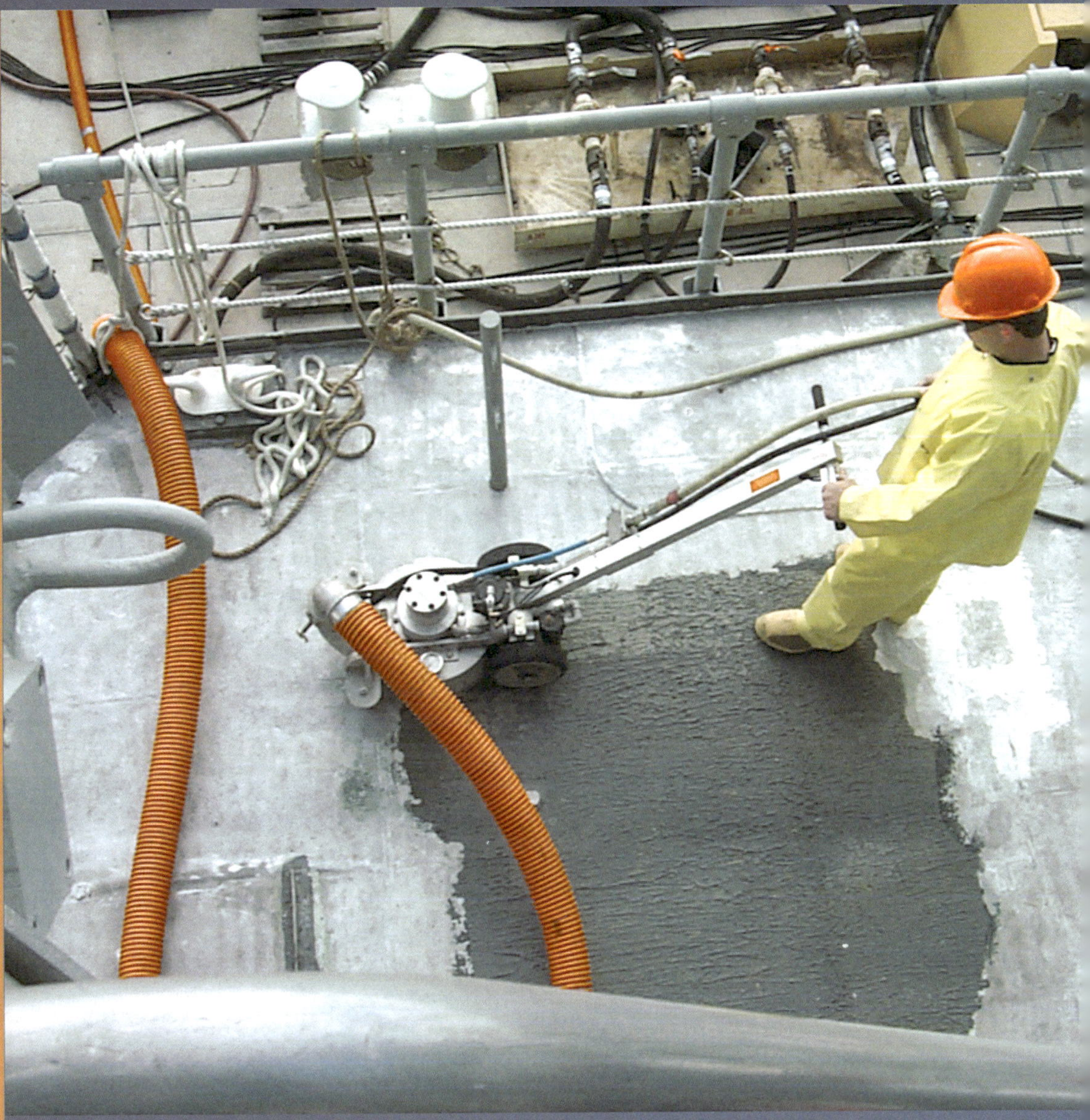

43101-12

Hydroblasting

Trainees with successful module completions may be eligible for credentialing through NCCER's National Registry. To learn more, go to **www.nccer.org** or contact us at **1.888.622.3720**. Our website has information on the latest product releases and training, as well as online versions of our *Cornerstone* newsletter and Pearson's product catalog.

Your feedback is welcome. You may email your comments to **curriculum@nccer.org,** send general comments and inquiries to **info@nccer.org**, or fill in the User Update form at the back of this module.

Objectives

When you have completed this module, you will be able to do the following:

1. Know the proper safety procedures for hydroblasting.
2. Explain the requirements for job-site setup for hydroblasting.
3. Perform the basic hookup for cleaning with a hydroblaster.
4. Perform job-site maintenance on dump valves and lances.
5. Identify types of hydroblasting equipment.

Performance Tasks

Under the supervision of your instructor, you should be able to do the following:

1. Explain the proper safety procedures for hydroblasting.
2. Describe the requirements for job-site setup for hydroblasting, including assembly of all hoses, valves, lances, and equipment.
3. Demonstrate the correct and effective use of flex and rigid lances, line moles, shotguns, and accessories.
4. Demonstrate the proper care of equipment and proper job site housekeeping and safety.
5. Identify types of hydroblasting equipment ans its uses, including 20,000 psi equipment.

Trade Terms

Antiwithdrawal device
Atmospheric hazard
Automation
Bypass valve
Capacity
Cavitation
Chemical cleaning
Chilblains
Cold stress
Combustible
Confined space
Discharge
Dump system
Emergency shutdown
Energy isolation
Fan jet
Fan tip
Filtration
Flame-retardant clothing
Flex lance
Flow rate

Flushing
Frostbite
Heat cramps
Heat exhaustion
Heat rash
Heat stress
Heatstroke
Hydraulic lance
Hydroblasting technician
Hydroexcavation
Hypothermia
Jet
Job safety analysis (JSA)
Lance
Lanyard
Line mole
Maximum allowable
    working pressure
    (MAWP)
Moling

Nonpermit-required
    confined space
Nozzle
Orifice
Oxygen-deficient
    atmosphere
Oxygen-enriched
    atmosphere
Permit-required confined
    space
Plug
Plunger
Positive displacement
Pressure rating
Pump operator
Quick-connect coupling
Radiation exposure badge
Rupture disk
Safety card
Safety shroud
Scaffold

Shielding
Shotgun
Spin nozzle
Stinger
Straight jet
Straight tip
Suction
Team leader
Test pressure (TP)
Throttling
Tip
Trench foot
Turbulence
Ultrahigh pressure (UHP)
Velocity
Water jet
Weep hole
Whip check
Working pressure (WP)

Industry Recognized Credentials

If you're training through an NCCER-accredited sponsor you may be eligible for credentials from NCCER's Registry. The ID number for this module is 43101-12. Note that this module may have been used in other NCCER curricula and may apply to other level completions. Contact NCCER's Registry at 888.622.3720 or go to nccer.org for more information.

Contents

Topics to be presented in this module include:

Figures and Tables

1.0.0 INTRODUCTION

Hydroblasting (*Figure 1*) is the process of cleaning surfaces with high-pressure water. The hydroblasting process uses pumps capable of producing pressures from around 1,000 to 48,000 pounds per square inch (psi). This water is then fed through a hose to a control valve, which directs the flow to the lance. The lance has a very small diameter nozzle leading to an orifice at its tip. This narrow nozzle releases the water at a very high velocity, sometimes so high that it can strip paint or boiler scale from metal. The same technology is used to dig holes in the ground, clean concrete, cut away decayed concrete, and even, when abrasives are mixed into the water, cut metal plate.

Hydroblasting and water-jetting equipment are very powerful. The standard type of system commonly used for cleaning is rated at 3,000 psi. This is the cleaning level, and the components rated for 15,000 psi are called standard components, as opposed to high-pressure components. To bring that pressure into perspective, the equivalent of 15,000 psi would be to place an object weighing 2½ to 7½ tons on an area of 1 square inch.

Pressure above 15,000 psi is generally considered ultrahigh pressure (UHP). (Some companies consider anything above 10,000 psi to be ultrahigh pressure.) Ultrahigh pressures are used to cut away coatings, including rubber tank linings and ships' hull paint. The special tips used at these pressures may include industrial sapphire tips. The equipment used at these pressure levels must be rated at the higher level. Hydroblasting and ultrahigh-pressure hydroblasting equipment must be kept separate.

WARNING!

Only thoroughly trained personnel should ever be allowed to operate the lance. This equipment can cut through heavy timbers and concrete block. Accidental contact with the jet is always a medical emergency. It is possible for skin penetration to take place at 100 psi. Even low-pressure systems running below 1,000 psi can seriously injure a person struck by the jet.

1.1.0 Theoretical Information

In order to understand the hydroblasting process, you need to have an understanding of its underlying theory. Fluids have certain characteristics that can be controlled to produce predictable results. The characteristics are pressure, flow rate, and velocity. Pressure (P) is a function of force (F) and area (A), calculated as follows:

$$P = F \times A$$

The factors that determine flow rate are volume, usually expressed in gallons, and time, usually expressed in minutes.

Velocity is calculated using the following equation:

$$\text{Velocity} = \text{Function} \times \text{Flow}$$

One unit of horsepower (hp) is equivalent to 746 watts. Small-horsepower pumps range from 60 to 200 hp. Horsepower is calculated using the following equation:

$$1 \text{ hp} = \frac{2\,\pi \times (\text{force} \times \text{radius}) \times \text{rpm}}{33,000}$$

The pressure level used is determined by the amount and type of material to be scoured from

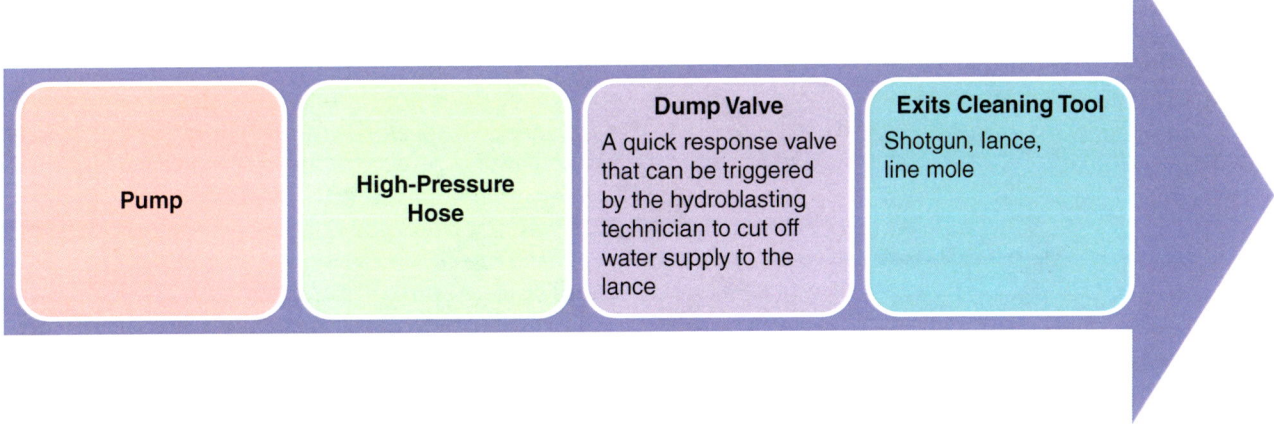

Figure 1 Hydroblasting.

43101-12-F01.EPS

surfaces, and the volume of water used depends on the amount of dirt and debris that needs to be removed during the hydroblasting process.

1.2.0 Pressure

Pressure is expressed either as pounds per square inch (psi), or as bars. In many countries, the metric system refers to kilograms per centimeter squared (kg/cm^2). Pounds per square inch is the English unit for the amount of push exerted against a surface (*Appendix A* has typical conversions between English and metric units). Hydroblasting uses pressure from approximately 1,000 psi to about 48,000 psi.

1.3.0 Flow Rate

Liquids flow to fit the shape of their container. The liquid itself does not change volume; this quality is called incompressibility with pressure. A pump causes the liquid to flow in a certain direction at a speed called the flow rate. Flow rate is the volume of the fluid that passes a certain point in a specific amount of time, usually expressed in gallons per minute (gpm), or liters per minute (lpm). Flow is faster if it is smooth, as opposed to turbulent. It is the need for smooth flow that requires the hose and pipe to have smooth interior walls; rough surfaces, like streambeds, make for turbulence, and decrease flow rates. In hydroblasting, the major controlling factor for flow rate at a given pressure is the size of the nozzle aperture.

1.4.0 Impact

Impact—how hard the water strikes the target—is related to the mass or weight of the water and the velocity of the stream; that is, the higher the velocity for a given amount of water, the greater the impact. The narrowing of the nozzle produces a very high-speed and high-impact stream striking a small area. This is also true of the back pressure from the lance. At a normal working pressure (WP) of 10,000 psi, the maximum reasonable flow rate would be about 8 gpm to avoid insupportable back pressure levels.

2.0.0 INDUSTRIAL AND LEGAL CAUTIONS

In industrial work, it is not possible to eliminate all hazards. It is, however, the responsibility of workers to reduce those hazards as much as possible. It is also the personal responsibility of human beings to reduce hazards at all times. The hazards associated with hydroblasting are very manageable if safe procedures are followed and diligence is a habit. Remember, nobody benefits from accidental injuries, so pay attention to what you and others are doing. Learn to do things the right way, and keep doing your work the right way.

One of the expressions of the concern for safety in the hydroblasting industry is the minimum age for performing hydroblasting work. You must be at least 18 years old to operate hydroblasting equipment.

2.1.0 Basic Hydroblasting Safety Rules

The full list of the Greater Baton Rouge Industry Alliance (GBRIA) Safety Fundamentals for Hydroblasting, adopted in 2011, is given in *Appendix B*. The following are some basic hydroblasting safety rules:

- Never point the nozzle of the lance at anybody, or allow the nozzle to point at your body. Remember, the water that comes out of the nozzle is capable of removing skin, and if it is close enough, it may cut through your flesh. Even with personal protective clothing, the jet can cut away the armor much more quickly than you would expect.
- Wear waterproof, safety-toed boots with metatarsal covering.
- Ensure toxic surface safety by knowing all the information on the safety data sheet (SDS).
- Be aware of the four points of entry of contaminating and infectious substances.
- Wear a protective face shield.
- Never point the lance at your body, whether you are wearing armor or not. Even the best armor is designed to resist a momentary swipe, not sustained streams of high-pressure water. The momentary carelessness that leads to trying to use the lance to clean your shoes or your face shield has immediate and very unpleasant consequences, even at relatively low pressures. Always be aware of where the lance is pointed, and always be aware of the hazards.
- If water does not divert when the dump valve is released, stop work immediately. Shut off pressure at the pump, drain the system, and have the valve repaired or replaced.
- Never try to repair components while they are pressurized. Verify with your supervisor that all stored energy has been released. Always stop the operation, shut down the pump, and release the pressure before trying to repair leaks or failures. Since all problems involve shutting off the pump, it must be obvious that the pump must, under no circumstances, be left unattended while it is running.

- If you see someone performing an unsafe activity, tell them to immediately stop.

2.1.1 Personal Protective Equipment

In addition to the standard equipment, all workers in the area of a hydroblasting operation must wear waterproof suits, head protection, and protective goggles or a full-face shield. Hydroblasting technicians must wear, at the least, protective metatarsal and shin guards, gloves, and a full-face shield. Since water and the material you are cleaning off is spraying around the area, it is a good idea to pause at intervals and wipe your face shield, so that you can see hazards, as well as seeing the work itself. Lance operators are not required to wear shin and metatarsal guards, only the required rubber boots. The waterproof clothing provides splash protection from water and contamination but is not strong enough to stop a direct swipe from a jet (*Figures 2 and 3*). Armor is available that has been manufactured to protect hydroblasting technicians from brief contact with the jet (*Figure 4*), but you should be aware that it only protects you from very brief contact with the jet. You should also be aware that the jet delivers a very heavy physical blow, even through the armor. You must be very careful of your footing, as the back pressure from the lance can cause you to fall if your foot slips. Depending upon the environment where the work is being performed, technicians may be required to wear flame-retardant clothing.

The sound level of a hydroblasting operation frequently runs in excess of 85 decibels, so hearing protectors or earplugs are required. In addition, if a hydroblasting technician is blasting possibly toxic substances from surfaces, breathing equipment or masks may be necessary.

2.2.0 Environmental Issues

It is very important that you know what is to be blasted, whether it is potentially toxic, and that the project is set up to deal with the wastewater being produced. While hydroblasting is a rela-

43101-12_F03.EPS

Figure 3 Waterproof clothing.

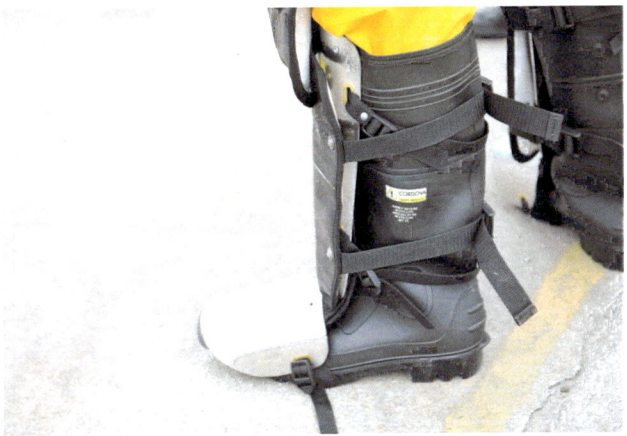

43101-12_F02.EPS

Figure 2 Waterproof foot protection.

43101-12_F04.EPS

Figure 4 Turtle skin personal protective equipment (PPE).

tively clean technology, as well as relatively efficient in its use of water, it still involves spraying some quantities of water over a period of time. In many contexts, water may be channeled to storm drain systems. In some applications, and in some particular locations, water is recovered, cleaned, and recycled, by using either filtration systems or a series of settling tanks or both. Be sure to check local codes and company requirements affecting discharge of cleaning water before setting up equipment for a project. Additional training may be required to collect and package waste that may be deemed hazardous under the Resource Conservation and Recovery Act.

Many times, the hydroblasting technician will be working to remove potentially toxic materials from vessels, surfaces, or pipes. The materials may be blasted loose and either thrown into the air as tiny particles or as vapors mixed with water. Be sure that you know what materials you may be releasing. If it is possible that you will be releasing toxic airborne contaminants, you must wear a breathing mask capable of protecting your lungs from the hazards.

The use of blast bags has a number of advantages. Blast bags are an environmentally friendly, safety enhancing, and cost-effective way to contain the majority of wastewater and residual product. Blast bags contain overspray, thus protecting surrounding equipment. They enhance safety by protecting workers from flying debris and high-pressure spray and reduce exposure to chemicals. Blast bags are cost-effective because they eliminate the cost of expensive cleanup, save money that would be spent on building and dismantling expensive backdrops, pans, and sanded areas, among other equipment, and they reduce downtime.

2.3.0 Accidents

A waterproof safety card should be posted in all hydroblasting work areas (*Figure 5*). Any and all blast accidents must be reported immediately. Even if the blast did not visibly penetrate the skin, any contact can force potentially contaminated water and entrained air bubbles and toxins under the skin. Even if there is no blood showing, there can be serious risk of infection and permanent damage, so take the injured person to a doctor, hospital, or emergency room immediately. The Water Jetting Trade Association says, in its *Recommended Practices for the Use of Manually Operated High-Pressure Water Jetting Equipment*, that all hydroblasting technicians should carry a waterproof medical alert card (*Figure 6*) to show to physicians that reads:

This person has been involved with high-pressure water jetting at pressures up to 14,500 lb/in² (100 MPa, 1,000 bar, 1,019 kg/cm²) with a jet velocity of 900 miles (1,440 km) per hour. Please take this into account when making your diagnosis. Unusual infections with microaerophilic organisms occurring at lower temperatures have been reported. These may be gram-negative pathogens such as are found in sewage. Bacterial swab and blood cultures may therefore be helpful.

If the incident takes place in a remote area, the wound should be covered, and the worker observed until medical care can be provided. Check your company's rules on accident reporting, and if any accident occurs, follow procedure. Report all accidents, no matter how small, to the appropriate supervisor.

2.3.1 Infection Control

Extreme caution and strict compliance with procedures must be used to prevent the hydroblast water jet stream from striking the operator or other employees. No portion of the body must ever be placed in front of the water jet. These jets of water can easily puncture and tear the skin or penetrate deeper, causing infection. Contaminants carried by the water can make even the smallest cuts susceptible to infection. The cleaning solution used may be absorbed into the bloodstream, which could be fatal. If possible, only clean water should be used. Raw water may contain small quantities of nitrogen, ammonia, or chlorine, which could be deadly if vaporized in a confined space.

WARNING!!!

An injury caused by high pressure waterjets can be serious. In the event of any waterjet injury:

— Seek medical attention immediately. Do not delay!

— Inform the doctor of the cause of the injury.

— Show the doctor this card.

— Tell the physician what type of waterjet project was being performed at the time of the accident and the source of the water.

IMPORTANT MEDICAL INFORMATION!

READ THIS PLASTIC CARD AND KEEP IT IN YOUR WALLET. IN THE EVENT OF A WATERJET INJURY, SHOW THE CARD TO YOUR DOCTOR.

Distributed by: WaterJet Technology Assn.-
Industrial & Municipal Cleaning Assn.,
906 Olive Street, Suite 1200,
St. Louis, MO 63101-1448,
phone: (314)241-1445, fax: (314)241-1449,
e-mail: wjta-imca@wjta.org,
website: www.wjta.org

43101-12_F05.EPS

Figure 5 Waterproof safety card.

2.3.2 Channeling

One of the injuries that can happen as a consequence of encountering water jets is called channeling. This is an injury where the jet enters the body through a small entrance wound, and then travels along some channel inside, to produce trauma or infection at some other place. Take no chances with injuries.

2.4.0 Indoor Engine Exhaust

If circumstances force you to share an enclosed space with the pump rig, you must provide effective exhaust arrangements to get the engine waste out of the area. It is also absolutely vital that you use hearing protection, as the sound in a restricted space is amplified. It is preferable that the pump be outside, if at all possible. If necessary, it is possible to rig an external exhaust system using a flex-tubing extension on the pump exhaust and a surrounding screen of tarpaulin. Whatever system you use, be sure to obtain permission from your supervisor or engineer.

2.5.0 Back Pressure

Another safety consideration is back pressure, which is the back thrust of the gun as the water flows. *Figure 7* shows the braced position of a worker at 4,000 psi. The factors in the equation affecting back pressure are pressure level and flow rate. At 10,000 psi, as a general rule, 8 gallons per minute would be a reasonable maximum flow rate. This pressure and flow rate would require a #6 nozzle with an aperture of 0.059 inches. At 36,000 psi, about 3 gpm would be a reasonable maximum, using a 0.028 aperture. Some workers can handle more back pressure, but fatigue also becomes an issue more quickly as back pressure increases.

Back pressure can be calculated using the formula: back pressure (B, in pounds) = 0.052 × flow (F) in gallons per minute (gpm) × square root of pressure (P) in pounds per square inch (psi). Using this formula, it is easy to calculate the back pressure from a shotgun. If you are planning a job where you are to use 10,000 psi and a flow of eight gallons per minute, the calculation becomes:

$$B = 0.052 \times F \times P$$
$$B = 0.052 \times 8 \times 10{,}000$$
$$B = 0.052 \times 8 \times 100 = 41.6 \text{ pounds}$$

This means that the technician operating the gun is resisting 41.6 pounds of pressure pushing into the technician's shoulder all the time. Obviously, that will produce fatigue after several hours. That also makes it clear why the technicians need to observe each other for signs of fatigue. You should keep track of these calculations when operating, as it will allow you to determine the length of shift for each technician. Remember that the back pressure changes as the flow rate and the pressure vary.

MEDICAL ALERT NOTE TO PHYSICIAN

This patient may be suffering from a waterjet injury. Evaluation and management should parallel that of a gunshot injury. The external manifestations of the injury cannot be used to predict the extent of internal damage. Initial management should include stabilization and a thorough neurovascular examination. X-rays can be used to assess subcutaneous air and foreign bodies distant from the site of injury. Injuries to the extremities can involve extensive nerve, muscle, vessel damage, as well as cause a distal compartment syndrome. Injuries to the torso can involve internal organ damage. Surgical consultation should be obtained. Aggressive irrigation and debridement is recommended. Surgical decompression and exploration may also be necessary. Angiographic studies are recommended pre-operatively if arterial injury is suspected. Bandages with a hygroscopic solution ($MgSO_4$) and hyperbaric oxygen treatment have been used as adjunctive therapy to decrease pain, edema and subcutaneous emphysema. Unusual infections with uncommon organisms in immunocompetent patients have been seen; the source of the water is important in deciding on initial, empiric antibiotic treatment, and broad-spectrum intravenous antibiotics should be administered. Cultures should be obtained.

43101-12_F06.EPS

Figure 6 Waterproof medical alert card.

43101-12-F07.EPS

Figure 7 Braced for back pressure.

2.6.0 Noise

The sound levels of hydroblasting operations are normally in excess of 85 decibels (*Figure 8*). At 85 decibels, the US Occupational Safety and Health Administration (OSHA) guidelines require single-protection hearing earphones or muffs over both ears for anyone exposed to sound at this level or above for eight hours a day. As the sound gets louder, which it does in hydroblasting work under different circumstances, the time of permissible exposure grows shorter. Always wear hearing protection of the approved types, either hearing protective earphones or approved earmuffs. Do not use substitutes such as cotton. Foam earplugs must be clean and sterile. Use new ones each time or use reusable plugs. Clean reusable plugs with dry hands.

At noise levels above 105 decibels, wear double-protection earphones and muffs over both ears. You must use hearing protection at all times while operating equipment. Follow your company's rules on hearing protection.

3.0.0 CONFINED SPACES

> **WARNING!**
>
> All permit-required confined spaces must be identified, and the associated permit must be posted. No one may enter a confined space without proper training and a valid entry permit. Some confined spaces require both a permit to enter and a permit to work in the space. Never enter a confined space if you are not sure how safe it is.

In hydroblasting work, one of the classes of hazard most commonly encountered is confined space work to prevent the need for vessel entry. High-pressure water is frequently used to clean tanks and chemical reaction vessels of various kinds, and the possibility always exists of encountering toxic atmospheres. Hydroblasting is also used for clearing sanitary sewers and inside sewer manholes, risking exposure to toxic atmospheres. Do not enter spaces that may have had toxic materials inside them until you have contacted the required supervisors and obtained a permit (*Figure 9*).

Spaces on a job site are considered confined when their size and shape restrict the movement of anyone who must enter, work in, and exit the space. Confined spaces often have poor ventilation and are difficult to enter and exit. For example, employees who work in process vessels generally must squeeze in and out through narrow openings and perform their tasks while in a cramped or awkward position. In some cases, confinement itself creates a hazard. Always ensure that employees are comfortable in confined spaces.

Permissible Noise Exposures*

Duration per Day, Hours	Sound Level dBA Slow Response
8	90
6	92
4	95
3	97
2	100
1 1/2	102
1	105
1/2	110
1/4 or less	115

* When the daily noise exposure is composed of two or more periods of noise exposure of different levels, their combined effect should be considered, rather than the individual effect of each. Exposure to impulsive or impact noise should not exceed 140dB peak sound-pressure level.

43101-12_F08.EPS

Figure 8 Noise level chart.

Attachment 16 – 2
Confined-Space Entry Permit

Master Card / Safe Work Ticket No. _____

1. Work Description: _____
    Equip. Name / Number & Location or Area _____
    Purpose of Entry _____
    Valid Start Date _____   Duration Time _____ to _____

2. Hazardous Materials:
    What did the equipment last contain? _____
    Will the work generate a hazardous atmosphere? ☐ Yes ☐ No If yes, specify hazards and controls.
    _____

3. Rescue Requirements:
    ☐ External, by attendant   ☐ Complex Rescue, by rescue team at point of entry
    ☐ Non-IDLH and/or Simple Rescue, by rescue team on-site   ☐ IDLH, by rescue team at point of entry
    Has the rescue team been notified of the entry? ☐ Yes   ☐ N/A Time of notification _____
    How will the rescue team be summoned for an emergency? ☐ Radio Channel: _____ ☐ Other: _____

4. Gas Test Requirements:
    LEL/O_2 – Instrument Mfg./No. _____/_____   Bump Check Time/Gas Tester _____/_____
    Toxicity – Instrument Mfg./No. _____/_____   Bump Check Time/Gas Tester _____/_____
    Frequency of Testing: ☐ Continuous* ☐ Other – Specify _____
    *Continuous monitoring results must be recorded every three hours.

Acceptable Levels	Results								
Oxygen: 19.5% – 23.5%									
Combustible Gas: %LEL <10%									
Other _____ < PEL† _____									
Other _____ < PEL† _____									
Other _____ < PEL† _____									

†Entry in excess of the PEL will require appropriate PPE.

5. Ventilation / Exhaust Equipment:
    ☐ None Required, Natural Ventilation Adequate   ☐ Forced Air Ventilation   ☐ Exhaust Ventilation
    Equipment Type: ☐ Air Powered Horn ☐ Electric Blower   Volume Required: _____ cfm

6. Personal Protection:
    ☐ Gloves (type) _____   ☐ Respirator-Type _____
    ☐ Goggles or Face Shield   ☐ Self-Contained Breathing Equipment
    ☐ Lifelines Attached to Harness   ☐ Other – Specify: _____
    ☐ Chemical Resistant Suit, Specify Type _____

7. Fire Protection: ☐ None Required ☐ Portable Fire Extinguisher – Type and Size:_____
    ☐ Fire Watch   ☐ Other – Specify: _____

43101-12_F09A.EPS

Figure 9 Confined-space entry permit form. (1 of 2)

8. Condition of Area and Equipment:

Required Yes	N/A		THESE KEY POINTS MUST BE CHECKED
		a.	Equipment locked and tagged out?
		b.	Piping is disconnected, capped or plugged, and/or blinded?
		c.	Equipment emptied, washed, purged & ventilated?
		d.	Low voltage or GFCI-protected equipment provided?
		e.	Explosionproof electrical equipment provided?
		f.	Provisions are made to barricade or post signs at entry points when attendant is not on duty.

Other Requirements:

9. Special Instructions: ☐ None   ☐ Check with issuer before starting work.

10. Approval

	Permit			Permit Acceptance		
	Supt. / Area Supv.	Date	Time	Maint. Supv. / Engineer / Contractor Supv.	Date	Time
Issued by						
Endorsed by						
Endorsed by						

11. Individual Review / Entrant Roster: I have been instructed in the proper work permit, confined-space entry, lockout/tagout procedures, and associated physical and atmospheric hazards, and have reviewed the gas testing results.

Entrants	Date	Time In / Out	Time In / Out	Time In / Out	Time In / Out	Time In / Out

I have been informed of the duties and responsibilities for the attendant, and the associated physical and atmospheric hazards, and have reviewed the gas testing results.

Attendants	Date	Time On / Off	Time On / Off	Time On / Off

12. Job Completion:
☐ Yes ☐ N/A Has the rescue team been notified?
☐ Yes ☐ No Is the work on equipment complete & the confined space ready to return to service?
☐ Yes ☐ No Has the worksite been cleaned and made safe?
Workers answering above questions: _____

13. Post Job Review: Were any hazards encountered or created during entry operations?
☐ Yes ☐ No If yes, describe: _____

Possible solutions: _____

Forward to job file within 7 days of job completion.

43101-12_F09B.EPS

Figure 9 Confined-space entry permit form. (2 of 2)

In a confined space, hazards such as poor air quality, toxins, explosions, fire, and moving machinery parts tend to be far more deadly. Consider the following information from the OSHA:

- From 1992–2005, there were 431 confined space incidents with 530 fatalities in the United States due to toxic and/or oxygen-deficient atmospheres.
- From August 18 to December 31, 2009, there were 36 fatalities and six hospitalizations for workers related to confined spaces.
- From January 1 to August 1, 2011, there were 22 fatalities and three hospitalizations for workers related to confined spaces.
- Almost 25 percent of fatalities related to confined spaces are attributed to repair/maintenance or cleaning/inspection activities.

Confined spaces may contain unknown hazards. In one instance, a worker was lowered into a 21-foot deep manhole on a looped chain seat. Twenty seconds after entering the manhole, he started gasping for air and fell. He landed face down in the water at the bottom of the manhole. An autopsy determined that he died from lack of oxygen.

Most confined spaces have restricted entrances and exits. Workers are often injured as they enter or exit through small doors and hatches. It can also be difficult to move around in a confined space and workers can be struck by moving equipment. Escapes and rescues are much more difficult in confined spaces than they are elsewhere.

Confined spaces are entered for inspection, equipment testing, repair, cleaning, or emergencies. They should only be entered for short periods of time. Confined spaces include the following:

- Manholes
- Boilers
- Trenches
- Tunnels
- Tubes
- Underground utility vaults
- Pipelines
- Pits
- Air ducts
- Process vessels (*Figure 10*)
- Railcars
- Vessel skirts
- Ship hulls

A written confined-space entry program can protect you. It identifies the hazards and specifies the equipment or support that is needed to avoid injury. All industrial and some construction sites have written confined-space entry programs. You must know and follow your company's policy.

3.1.0 Confined-Space Classification

Confined spaces must be inspected before work can begin. This helps to identify possible hazards. After an inspection by a company's authorized person, the confined space is classified based on any hazards that are present. The two classifications are nonpermit-required confined space and permit-required confined space. See *Figure 11.*

3.2.0 Nonpermit-Required Confined Space

A nonpermit-required confined space is a work space free of any mechanical, physical, electrical, and atmospheric hazards that can cause death or injury. After a space has been classified as nonpermit-required, workers can enter using the appropriate personal protective equipment for the type of work to be performed. Always check with your supervisor if it is unclear what personal protective equipment is required. A space is considered confined if it:

- Is large enough and so configured that an employee can bodily enter and perform assigned work

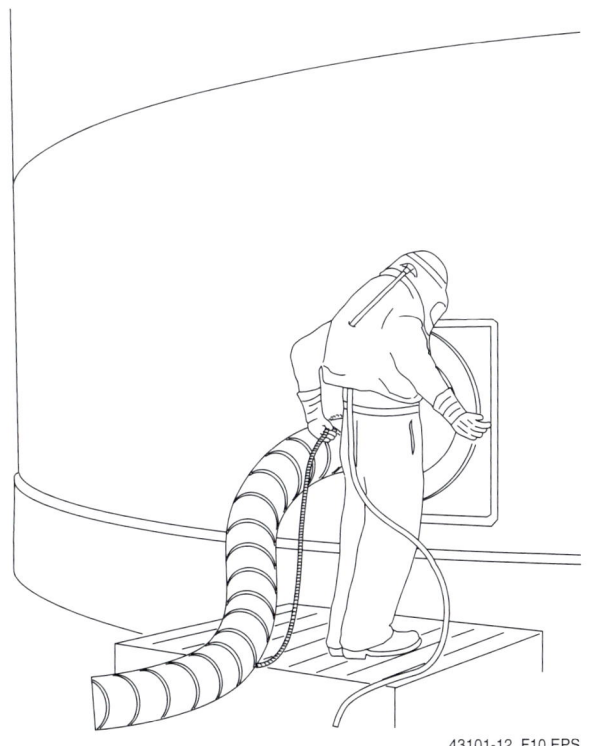

43101-12_F10.EPS

Figure 10 Entry into a process vessel.

- Has a limited or restricted means of entry or exit, such as tanks, vessels, silos, storage bins, hoppers, vaults, and pits
- Is not designed for continuous employee occupancy

An entry permit must be issued and signed by the job-site supervisor before the confined space is entered. No one is allowed to enter a confined space unless there is a valid entry permit. The permit is to be kept at the confined space while work is being performed. Always check with your supervisor if it is unclear whether or not you need a permit to enter a confined space.

3.3.0 Permit-Required Confined Space

A permit-required confined space is a confined space that has real or possible hazards. These hazards can be atmospheric, physical, electrical, or mechanical. *OSHA 29 CFR 1910.146* defines a permit-required confined space as a confined space that:

- Contains or has the potential to contain a hazardous atmosphere
- Contains a material that has the potential for engulfing an entrant (*Figure 12*)
- Has an internal configuration such that an entrant could be trapped or asphyxiated by inwardly converging walls or by a floor that slopes downward and tapers to a small cross-section
- Contains any other recognized serious safety or health hazard

3.4.0 Hazards

Confined spaces are dangerous. The main hazards include poor airflow and restricted movement. Poor ventilation can allow toxic gases to build up. Physical hazards are more dangerous because escape is difficult.

3.4.1 Atmospheric Hazards

Atmospheric hazards are the most common hazards in a confined space. In a hazardous atmosphere, the air can have either too little or too much oxygen, be explosive or flammable, or con-

NON-PERMIT REQUIRED CONFINED SPACE

PERMIT REQUIRED CONFINED SPACE

43101-12-F11.EPS

Figure 11 Classifications of confined spaces.

43101-12_F12.EPS

Figure 12 Engulfment.

tain toxic gases. Special meters are used to detect these atmospheric hazards (*Figure 13*).

3.4.2 Oxygen-Deficient or Oxygen-Enriched Atmospheres

A confined space that does not have enough oxygen is called an oxygen-deficient atmosphere; a confined space that has too much oxygen is called an oxygen-enriched atmosphere. For safe working conditions, the oxygen level in a confined space must range between 19.5 percent and 23.5 percent by volume, with 21 percent being con-

43101-12-F13.EPS

Figure 13 Detection meters.

sidered the normal level. Oxygen concentrations below 19.5 percent by volume are considered oxygen-deficient; those above 23.5 percent by volume are considered oxygen-enriched.

Many of the processes that occur in a confined space use oxygen and may reduce the percentage of oxygen to an unsafe level. These processes include the following:

- Burning
- Rusting of metal
- Breaking down of plants or garbage
- Oxygen mixing with other gases

> **WARNING!**
>
> Normal breathing in a confined space can also create an oxygen deficiency.

When the oxygen in a confined space is reduced, it becomes harder to breathe. The symptoms of insufficient oxygen happen in this order:

1. Fast breathing and heartbeat
2. Impaired mental judgment
3. Extreme emotional reaction
4. Unusual fatigue
5. Nausea and vomiting
6. Inability to move your body freely
7. Loss of consciousness
8. Death

Too much oxygen in a confined space is a fire hazard and can cause explosions. Materials like clothing and hair are highly flammable and burn rapidly in oxygen-enriched atmospheres. Fires can start easily in a confined space with oxygen-enriched air.

3.4.3 Combustible Atmospheres

Air in a confined space becomes combustible when chemicals or gases reach a certain concentration. Flammable gases can be trapped in confined spaces. These include acetylene, butane, propane, methane, and others. Dust and work by-products from spray painting or welding can also form a combustible atmosphere.

Some flammable gases are lighter than air and have a higher concentration at the top of a confined space. Vapors from fuels are generally heavier than air and form a greater concentration at the bottom of the space.

A spark or flame will cause an explosion in a combustible atmosphere.

3.4.4 Toxic Atmospheres

Toxic gases and vapors come from many sources. They can be deadly when they are inhaled or absorbed through the skin above certain concentration levels. In spaces with no ventilation, high concentrations can gather and quickly become toxic. Even in lower doses, some chemicals can seriously affect your breathing and brain functions.

The harmful effects of toxic gases and vapors vary. Many toxic gases, such as carbon monoxide, cannot be detected by sight or smell. Some toxic gases have harmful effects that may not show up until years after contact. Others, such as nitric oxide, can kill quickly.

NOTE

The material safety data sheet (MSDS) attached to the entry permit will have information about any toxins you may encounter.

3.4.5 Monitoring the Atmosphere in Confined Spaces

The air inside a permit-required confined space must be tested before anyone is allowed to enter. Atmospheric tests must be done in this order:

1. Oxygen content

2. Flammable gases and vapors

3. Potential toxic contaminants

The test for oxygen must be performed first because most combustible gas meters are oxygen-dependent and do not provide reliable readings in an oxygen-deficient atmosphere. The test for combustible gases is performed next because the threat of fire or explosion is usually more urgent and more life threatening than exposure to toxic gases and vapors.

Various instruments are used to test and monitor confined-space atmospheres. Portable, battery-operated gas detection meters can measure oxygen levels by changing the sensor in the detection meter. Gas detection meters must be calibrated and operated according to the manufacturer's instructions. The meter must be able to

detect oxygen and combustible gases at the levels specified in *OSHA 29 CFR 1910.146*. A radiation exposure badge may need to be worn by technicians to monitor any dose of radiation exposure presented in an environment.

3.5.0 Safeguards

It is important to understand how to protect yourself and your coworkers when working in a confined space. In order to do this, everyone must be aware of what is happening on the site and understand how to work safely. Energy isolation devices are often used so that equipment does not become active unexpectedly. These are the most common safeguards that should be used during confined-space operations:

- Monitoring and testing
- Ventilation
- Personal protective equipment
- Communication
- Training

3.5.1 Monitoring and Testing

The air in a confined space must be tested before any workers enter it. Testing must be done by a properly trained, qualified person. This person must be a company-approved or otherwise designated individual. He or she is referred to as the confined-space attendant. The air is tested for oxygen content, explosive gases or vapors, and toxic chemicals. This can be done by inserting a wand attached to a gas meter into the confined space.

Case History

Added Oxygen Causes Lost Life

A blaster went into a confined space to waterblast a tanker. The tanker had VCM (vinyl chloride monomer) that mixed with water and turned into carbon monoxide and displaced the oxygen. He died and the whole crew that followed him in died as well.

The Bottom Line: Make sure you know what you're going to clean, review the SDS, ensure the PPE is correct, and verify the rescue plan. Do not rescue the worker yourself. This accident could have been avoided. It happened because of poor communication between workers and unsafe work practices.

The atmosphere in a confined space may need to be monitored during the entire job. This is done by attaching monitors to entrants or by using outside devices. When the atmosphere is monitored, workers can be assured that the air quality is good and that they will immediately know about changes in the atmosphere that would require them to leave the space.

3.5.2 Ventilation

If the air in a confined space is hazardous or has the potential of becoming dangerous, the space must be ventilated immediately to remove toxic gases or vapors and replace lost oxygen. Ventilators blow clean air into the space, as shown in *Figure 14*. They must stay on as long as workers are in the space.

It is important to remember that just because there is air being blown into or being removed from the space, it doesn't mean that the air is being ventilated. Toxic gases can hide in confined spaces. Make sure the attendant has carefully tested the entire space before entering it.

3.5.3 Personal Protective Equipment

Every hydroblasting job requires some type of personal protective equipment. Standard personal protective equipment includes hard hats, safety goggles and glasses, boots, and gloves. On a confined-space job site, the following items may be needed in addition to the standard equipment:

- Full-body harness
- Lifelines
- Air-purifying respirator
- Air-supplying respirator

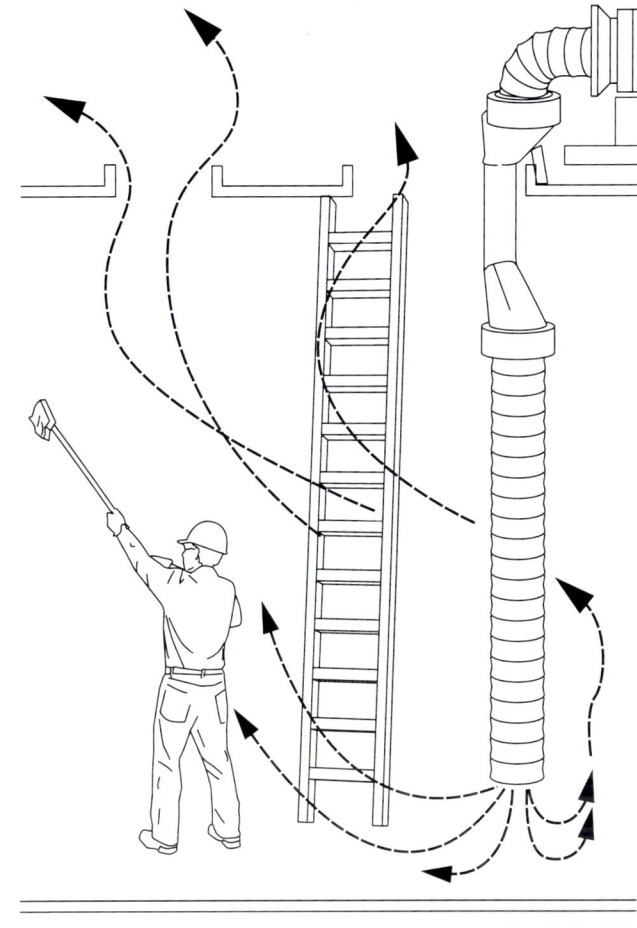

43101-12-F14.EPS

Figure 14 Proper positive ventilation.

Case History

Four Die in Confined Space

A project involved the upgrade/replacement of a sewer pumping station and the contractor prepared a confined-space entry permit for the work. One employee was disconnecting a sewer bypass connection in a manhole while three others were at the manhole entrance. The manhole filled with sewage and gases from the sewer line, and the employee was overcome by a lack of oxygen. The other three employees tried to help. Each entered the manhole one at a time, apparently to attempt rescue. Each was overcome by the sewer gases and died.

The Bottom Line: Asphyxiation (lack of oxygen) is not like what you see in the movies. You can't go in unprotected, stagger around awhile, save a couple of lives, then exit coughing and unharmed. Reality is quite different. Upon your first breath, you pass out. If a retrieval system is not in place, anyone who enters the space to rescue you will die. Many workers have died simply by putting their heads in manholes to assess a situation. Even without fully entering the confined space, these workers were immediately incapacitated.

3.5.4 Communication

All workers on a site must be able to communicate with one another. It is especially important for attendants and entrants to be able to communicate. This allows attendants to warn entrants about dangers and order an evacuation when necessary. Communication between workers is another way to monitor the confined space. For hydroblasting technicians, however, the noise level makes many normal communication systems unusable. Make sure that you establish signals that do not depend on audible communication before beginning work in a confined space.

3.5.5 Training

Entering a confined space requires specialized training. In fact, no one is allowed to enter a confined space unless they have been properly trained and authorized by the site supervisor to enter the space. Training gives workers the knowledge needed to complete their jobs safely and efficiently. If you have not been properly trained, do not enter any confined space.

4.0.0 SCAFFOLDS

Hydroblasting technicians use various types of motorized equipment throughout their careers. The back pressure from hydroblasting is too heavy to work from ladders or other unstable surfaces. If you are blasting a surface more than two feet above eye level, or about eight feet above the standing surface, the best thing to use is a scaffold.

4.1.0 Introduction to Scaffolds

Scaffolds consist of elevated working platforms that support workers and materials. They are a very common sight on many construction jobs. A typical scaffold is shown in *Figure 15*. The main part of the scaffold is the working platform. A working platform should have a guardrail system that includes a top rail, midrail, toeboard, and screening. To be safe and effective, the top rail should be approximately 42 inches high, the midrail should be located halfway between the toeboard and the top rail, and the toeboard should be a minimum of 4 inches high.

If people will be passing or working under the scaffold, the area between the top rail and the toeboard must be screened. Finally, the platform planks must be laid closely together. For safety purposes, the ends of the planks must overlap at least 6 inches and no more than 12 inches.

There are many types of scaffolds used in residential and commercial construction. The three most common types are built-up scaffolds, swing scaffolds, and beam-suspended scaffolds.

4.1.1 Scaffold Hazards

> **WARNING!**
>
> Only qualified scaffold personnel can modify scaffolds. This training does not qualify you for modifying scaffolds.

Improper or careless use of scaffolding can result in accidents, injury, or death. Those who work on scaffolds can minimize their risk by being aware of the hazards involved and following the proper safety procedures and guidelines to minimize those hazards. The main hazards involved with the use of scaffolds are:

- Falls
- Workers being struck by falling objects
- Electric shock

Falls can happen because fall protection has not been provided, is not used, or is installed or used improperly. Poorly planked scaffolding causes many falls. Working on scaffolds when conditions are dangerous, such as in high winds, ice, rain,

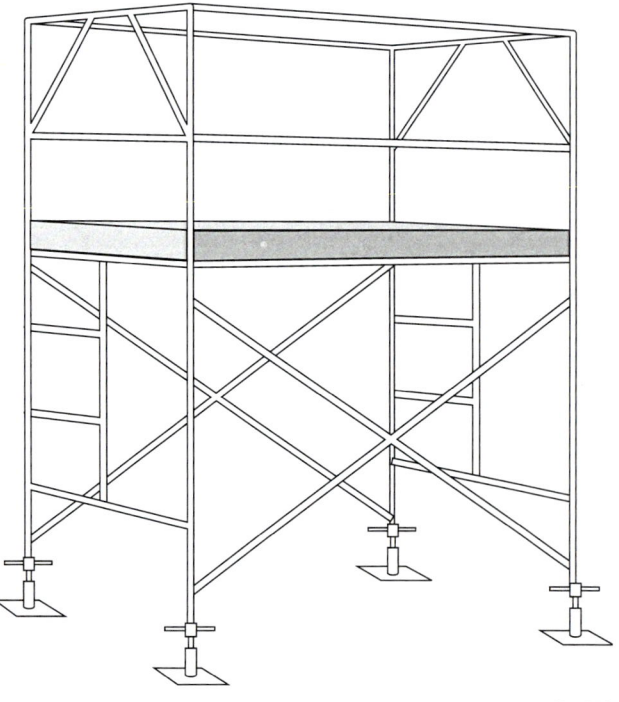

43101-12-F15.EPS

Figure 15 Typical scaffold.

---

and lightning, also leads to accidents. Falls also happen when scaffolding collapses because of improper construction.

Generally, fall protection is required on any scaffolds six feet or more above a lower level. Refer to OSHA standards or your company policy. Fall-protection devices consist of guardrail systems, personal fall-arrest systems, lanyards, and/or safety nets. A guardrail system normally serves as adequate fall protection for most scaffolding.

Guardrail systems must extend around all open sides of the scaffolding. The side facing the work surface need not have a guardrail if it is located less than 14 inches away from the work surface. Any opening on a scaffold platform must be protected by a guardrail system, including the access opening(s) and platforms that do not extend across the entire width of the scaffold.

People who work or pass under scaffolds may be hit by falling objects. Tools, materials, debris, and scaffolding parts may fall to the surface below. Those working on scaffolds may also be injured if there are others working above them, or if the structure or workpiece extends above the work level of the scaffold. Any worker who is exposed to the danger of falling objects is required to wear a hard hat. Depending on the situation, additional protection may be needed such as debris nets, screens or mesh, canopy structures, and toeboards. Barricades that prevent access under the scaffold can also be used to protect workers and others.

> **WARNING!**
>
> If the scaffold you are working on shifts or begins to collapse, stop what you are doing and exit the scaffold immediately.

> **WARNING!**
>
> Water and process chemicals on scaffold can create slippery surfaces. Decontaminate the surface before resuming work.

Because most scaffolding is made of metal, the chance of electric shock is always a hazard. Never assume that you can work around high-voltage wires just by avoiding contact. High voltages can arc through the air and cause electrocution without direct contact. When scaffolds must be erected close to power lines, the utility company must be called in to deenergize, move, and/or cover the lines with insulating protective barriers.

4.1.2 General Safety Guidelines

Safety begins by being trained in the proper use of scaffolds. It is equally important to always use the right safety equipment, including a hard hat and personal fall-protection systems.

Never work on scaffolds, in a work cage (*Figure 16*), or on a platform if you:

- Are subject to seizures
- Become dizzy or lightheaded when working at an elevation
- Take medication that might affect your stability and/or performance
- Are under the influence of drugs and/or alcohol

When working on scaffolds, always follow these guidelines:

- Erect and use scaffolds according to the manufacturer's instructions. They must also be erected and used in accordance with all local, state, and federal/OSHA requirements.
- Never interchange parts of a scaffolding system made by different manufacturers.
- Attach a complete (green), caution (yellow), or danger (red) tag (*Figure 17*) as needed to any scaffold that is assembled and erected to alert users of its current mechanical and/or safety status. Do not rely solely on the tag. Inspect all parts of scaffolding before each use.
- If the scaffold shifts, exit the scaffold immediately.

43101-12-F16.EPS

Figure 16 Work cage.

4.1.3 Safety Guidelines for Built-Up Scaffolds

Built-up scaffolds (*Figure 18*) are built from the ground up at a job site. Use the following guidelines when inspecting and using tubular built-up scaffolds:

- Inspect all scaffolding parts before assembly.
- Never use parts that are broken, damaged, or deteriorated. Be cautious of rusted materials.
- Follow the manufacturer's recommendations for the proper methods of erecting and using scaffolds. Do not make any modifications.
- Do not interchange parts from different manufacturers.
- Provide adequate sills or underpinnings for all scaffolding built on filled or soft ground. Compensate for uneven ground by using adjusting screws or leveling jacks.
- Do not use boxes, concrete blocks, bricks, or other similar objects to support scaffolds.

- Keep scaffolds free of clutter and slippery material.

- Be sure the scaffold is plumb and level at all times. Follow the prescribed spacing and positioning requirements for the parts of the scaffold. Anchor or tie scaffolds to the building at prescribed intervals.
- Use ladders rather than cross braces to climb the scaffold. Position ladders with caution to prevent the scaffold tower from tipping.
- Do not work on scaffolds that are more than 10 feet high without guardrails, midrails, and toeboards on open sides and ends.
- Lock the casters of mobile scaffolding when it is positioned for use.
- Do not ride on mobile scaffolds.
- Avoid building scaffolds near power lines.

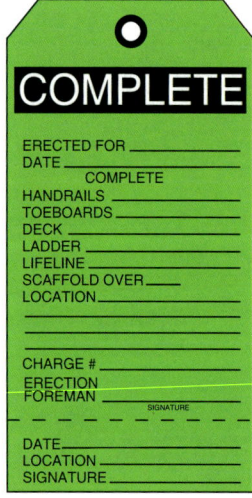
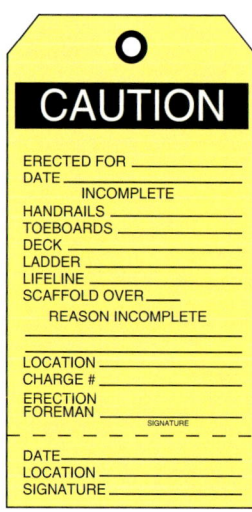

43101-12-F17.EPS

Figure 17 Typical scaffold tags.

Case History

Inspect All Materials

A crew laying bricks on the upper floor of a three-story building built a 6-foot platform to connect two scaffolds. The platform was correctly constructed of two 2-inch × 12-inch planks with standard guardrails. One of the planks, however, was not scaffold-grade lumber. It also had extensive dry rot in the center. When a bricklayer stepped on the plank, it broke and he fell 30 feet to his death.

The Bottom Line: Make sure that all planking is sound and secure. Your life depends on it.

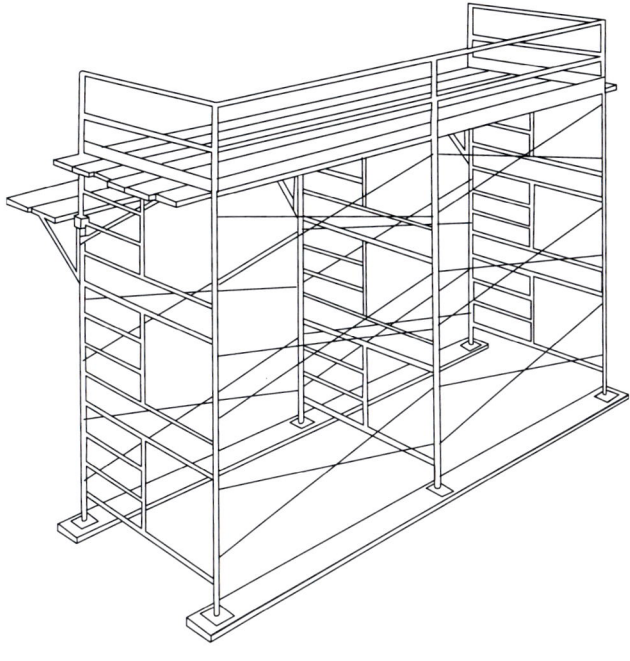

Figure 18 Built-up scaffold.

4.1.4 Safety Guidelines for Swing and Other Suspended Scaffolds

Swing scaffolds (*Figure 19*) are suspended by ropes or cables in a manner that allows it to be raised or lowered as needed. Another type of suspended scaffold is a work cage. A work cage is typically suspended with rigging devices that attach to I-beams with various sizes of clamps and rollers.

Follow the manufacturer's recommendations for installation, use, and maintenance of the equipment. Before installation, inspect all parts of a structure to which rigging and tieback lines will be secured to ensure that they can support the load.

4.2.0 Rescue After a Fall

Every elevated job site should have an established rescue and retrieval plan. Planning is especially important in remote areas that are not readily ac-

cessible. Before beginning work, make sure that you know what your employer's rescue plan calls for you to do in the event of a fall. Find out what rescue equipment is available and where it is located. Learn how to use equipment for self-rescue and the rescue of others. Ensure that you sit in the harness correctly to prevent blood pooling.

If a fall occurs, any employee hanging from the fall-arrest system must be rescued safely and quickly. Remember that you have a limited time to be rescued. Your employer should have previously determined the method of rescue for fall victims, which may include equipment that lets the victim rescue himself or herself, a system of rescue by coworkers, or a way to alert a trained rescue squad. If a rescue depends on calling for outside help such as the fire department or rescue squad, all the needed phone numbers must be posted in plain view at the work site. In the event a coworker falls, follow your employer's rescue plan. Call any special rescue service needed. Communicate with the victim and monitor him or her constantly during the rescue.

5.0.0 HYDROBLASTING EQUIPMENT

The most basic hydroblasting requirement is a high-pressure, positive-displacement pump (*Figure 20*), producing anywhere from 1,000 psi to 48,000 psi. The driver for the pump may be powered by a diesel- or gasoline-powered engine or an electric motor. The pump supplies water at pressure through a dump system, a quick-response valve that can be triggered by the hydroblasting technician to cut off water supply, to the lance. The lance may be of several varieties, including the flex lance, the rigid lance, the shotgun, or one of several crawlers or floor-blasting models. In each case, the water is delivered to one or more tips, jets, or nozzles, which narrow the flow to one or more apertures as small as six one-thousandths of an inch. Since the water is still flowing at the same rate when passing through the nozzle as when it passes through the larger lance tube,

---

Case History

Work Independently

Two workers were sandblasting a 110-foot-high water tank while working on a two-point suspension scaffold 60 to 70 feet above the ground. The scaffold attachment point failed, releasing the scaffold cables, and the scaffolding fell to the ground. Both workers died instantly because they were not tied off independently, nor was the scaffold equipped with an independent attachment system.

The Bottom Line: Make sure that all safety equipment is in place before using any scaffolding.

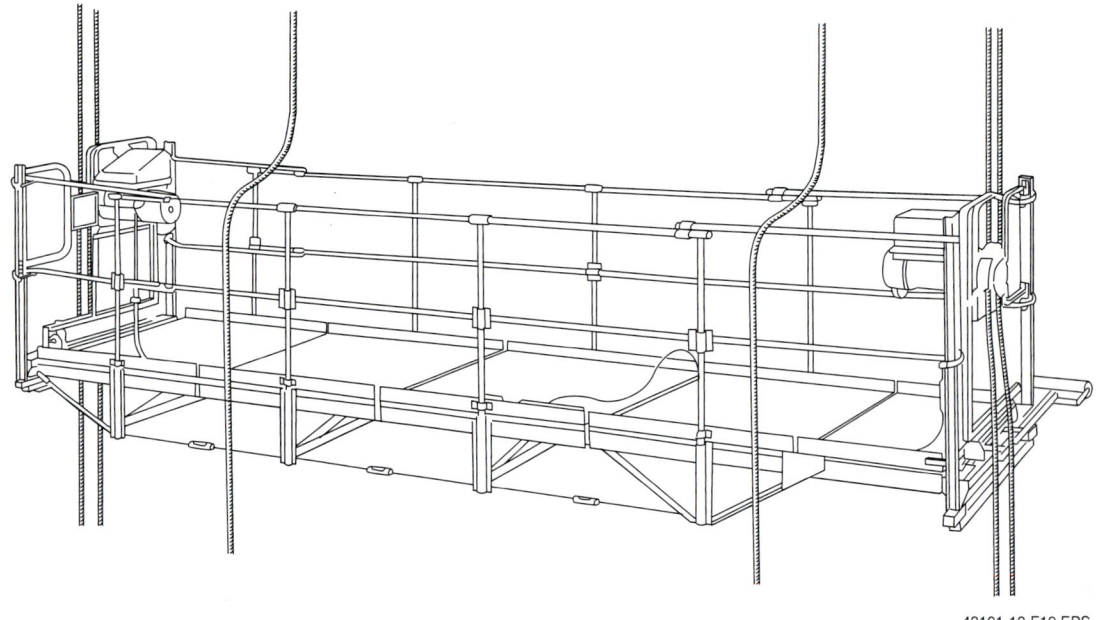

Figure 19 Swing scaffold.

it speeds up to get the same amount through in a given time. The nozzle may produce a very narrow cylindrical flow, called a straight jet (*Figure 21*), or a fan-shaped flow pattern, called a fan jet (*Figure 22*). In some applications, a vacuum system recovers and recycles the used water; in other cases, it is disposed of in storm drains, sanitary tubes, or held in retention ponds for later remediation.

Such power is present in the high-pressure water in hydroblasting that any weakness or malfunction becomes dangerous. With pressures ranging from 2½ tons to 20 tons on every square inch of the discharge side of the system, a pinhole leak is not a small problem for long. All equipment must be maintained at to the manufacturer's recommendations and inspected regularly.

5.1.0 Pumps and Valves

Pumps are used in almost every industry to move liquids and semisolids from one location to another. Pumps come in many types and sizes for various applications. They range from small metering pumps that pump only ounces per day to very large, high-capacity pumps that pump thousands of gallons per hour.

5.2.0 Pump Types

A pump is a mechanical device designed to increase the energy of a fluid so that a quantity of the fluid can be transported from one location to another. Pumps are generally driven by electric motors and are used to pump water, stock, coating, additives, oil, and hydraulic fluid.

All pumps operate under the general principle that fluid is drawn in at one end, the suction side, and forced out at the other end, the discharge side, at an increased velocity.

Pumps are classified as either positive displacement or centrifugal. Positive-displacement pumps are generally used for pumping liquids that have very high viscosities; that is, they are thick and flow slowly. Positive-displacement pumps are generally high-pressure pumps. Centrifugal pumps are used for pumping liquids that have low viscosities, such as water or thin stock. Centrifugal pumps are low-pressure pumps.

Figure 20 Hydroblasting pump.

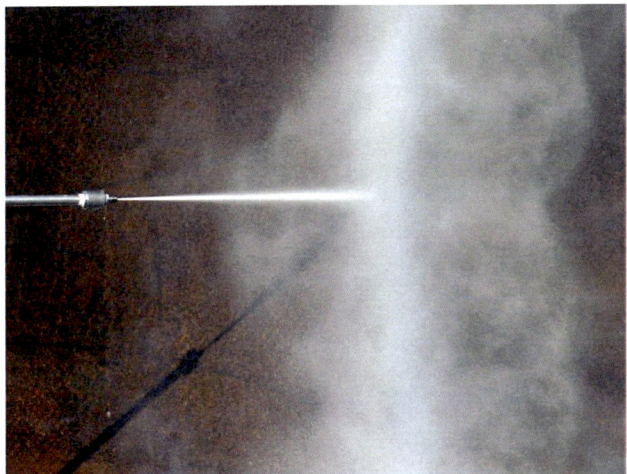

Figure 21 Straight jet.

43101-12-F21.EPS

43101-12-F22.EPS

Figure 22 Fan jet.

A positive-displacement pump takes a discrete amount of liquid and moves it physically from one place to another. A centrifugal pump adds kinetic energy to the liquid, and provides an enclosure that forces the liquid to go in a particular direction. Positive-displacement pumps produce a pattern of intermittent flow; the liquid flow is interspersed with intervals of no flow, although this can be so nearly continuous as to be undetectable. Centrifugal pumps are continuous-flow pumps.

Because positive-displacement pumps create very high pressures very quickly in a closed discharge pipe, there is usually a relief valve between the pump and the first closing point, such as a valve or possible pipe obstruction. In most hydroblasting applications, a bursting disk functions as an emergency relief valve.

The most common types of pumps in hydroblasting work are rotary pumps and reciprocating pumps.

5.3.0 Reciprocating Pumps

Reciprocating pumps operate by back-and-forth or up-and-down, straight-line motion. Reciprocating pumps require a suction, or intake, stroke and a discharge stroke to move the fluid.

During the suction stroke, a check valve in the suction line opens and a check valve in the discharge line closes. During the discharge stroke, the suction check valve closes and the discharge check valve opens. The action of the check valves prevents the liquid from flowing back out of the suction line on the discharge stroke. *Figure 23* shows check valves in the suction and discharge lines.

The pistons of reciprocating pumps create a back-and-forth movement to deliver a pulsating flow of fluid. This movement causes the flow to go from zero to maximum pressure and volume and then return to zero.

To compensate for this flow pulsation, an air chamber (*Figure 24*) is used in the discharge side of the reciprocating pumps. Air trapped in this chamber compresses as the pressure increases during the discharge stroke. The air expands as pressure decreases during the suction stroke. This reduces the extreme changes in pressure and results in a smoother fluid flow. The reciprocating pump moves liquid by displacing the liquid with a piston. This operating principle is called positive displacement. *Figure 25* shows an example of a positive displacement plunger pump. The most common types of reciprocating pumps generally used in hydroblasting are piston or plunger pumps.

5.3.1 Piston Pumps

In piston pumps (*Figure 26*), the pumping element is a piston, which is a short, cylindrical part that moves back and forth in the pump chamber, or cylinder. The stroke of the piston is usually greater than the piston length. Leakage of fluid around the outside of the piston is controlled by piston rings or packing.

5.3.2 Plunger Pumps

The plunger pump is often confused with the piston pump. The biggest difference is that a piston moves back and forth within a cylinder, while the plunger moves into and withdraws from a cylinder. *Figure 27* shows the difference between plungers and pistons in reciprocating pumps.

Unlike a piston, the length of a plunger is greater than its stroke. In the plunger pump, the packings in the cylinder control leakage around the outside of the plunger. *Figure 28* shows a plunger.

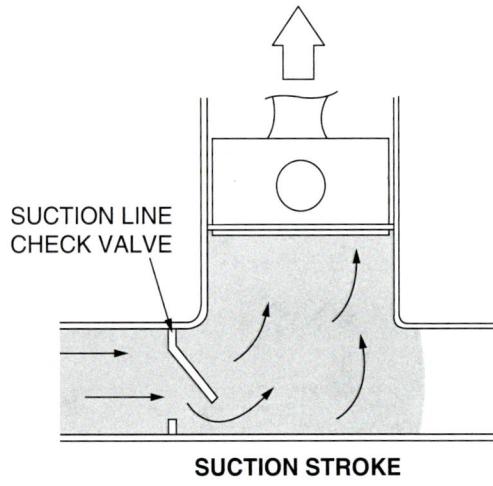

SUCTION STROKE

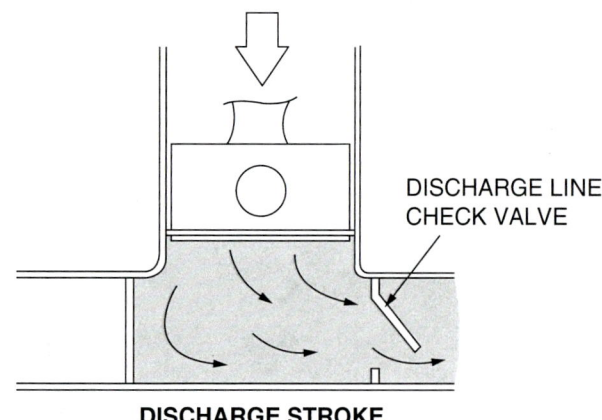

DISCHARGE STROKE

43101-12-F23.EPS

Figure 23 Check valves.

5.4.0 Valves

In hydroblasting, valves are devices that control the flow of fluids through a piping system. While the designs of valves vary, all valves have two common features: a passageway through which fluid flows and a moveable part that opens and closes the passageway. A valve can be used to provide on-off service only, can act as a throttling device to allow different flow rates, can relieve excess pressure in a pipeline, or can prevent reversal of flow through a line. There are basically four ways to control the flow through a piping system, and each type of valve uses one or more of these methods. These methods are as follows:

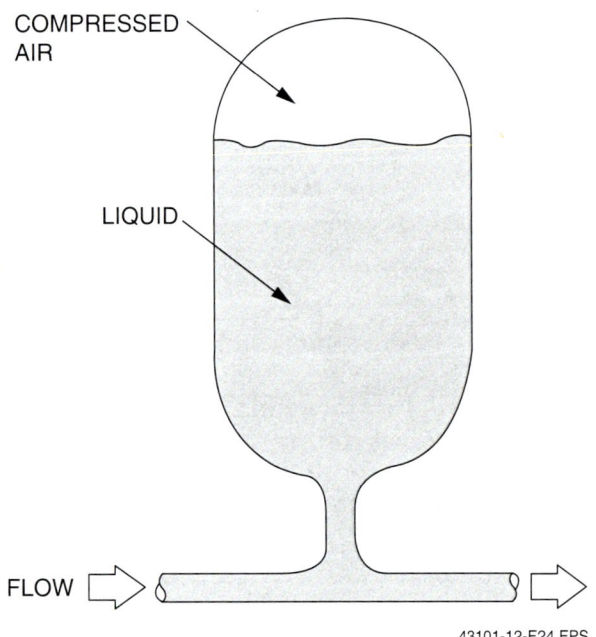

Figure 24 Air chamber.

- Moving a disk or plug into or against a passageway
- Sliding a flat, round surface across a passageway
- Rotating an opening inside a shaft across the passageway
- Moving a flexible material into the passageway

Valves are made of a variety of materials, including bronze, iron, carbon steel, aluminum, alloy steels, and PVC. The application of the valve usually dictates what type of material is used, and many times, there is more than one material that meets the requirements of an application. Bronze or iron with bronze trim are used for valves in air or water services. Iron is also used for valves in low-pressure steam services. Steel is used for valves that regulate noncorrosive products, and ductile iron, which is less expensive than steel, offers better resistance to mechanical or thermal shock than other metals provide.

Stainless steel is often used to regulate the flow of corrosive materials and in services that require

Case History

Fatality

A technician was working inside a confined space on scaffolding with a shotgun screen inside a paper mill vessel. The hydroblasting hose was not properly secured on the work platform. The weight of the hose and the effect of gravity pulled the shotgun from the operator's arm, wrapping around his feet and pulling him off the scaffold to his death.

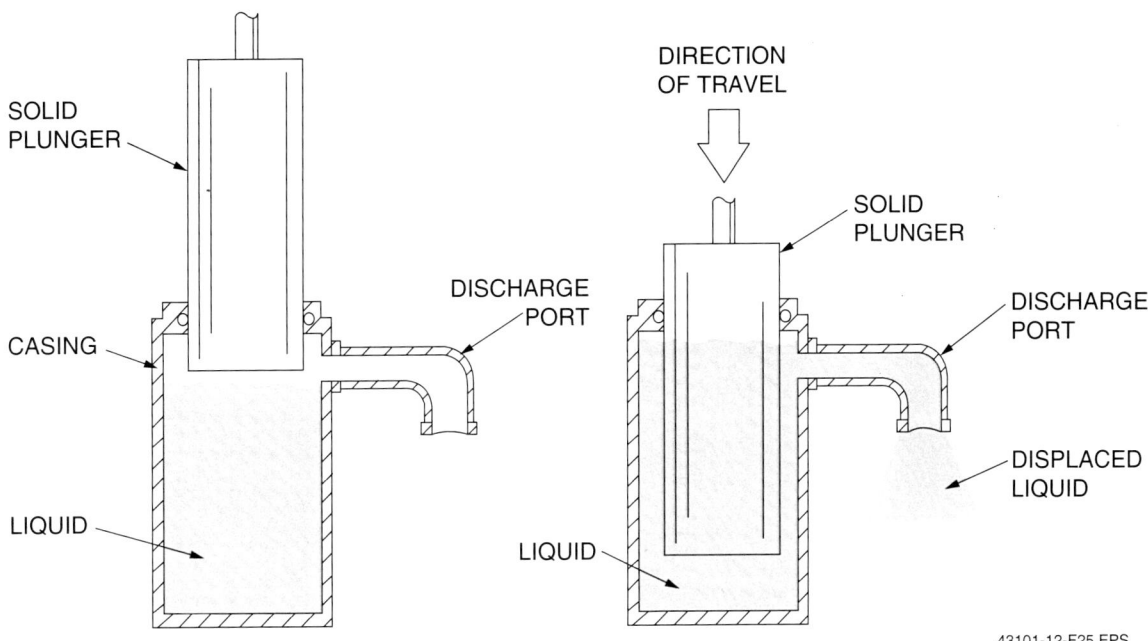

Figure 25 Positive displacement.

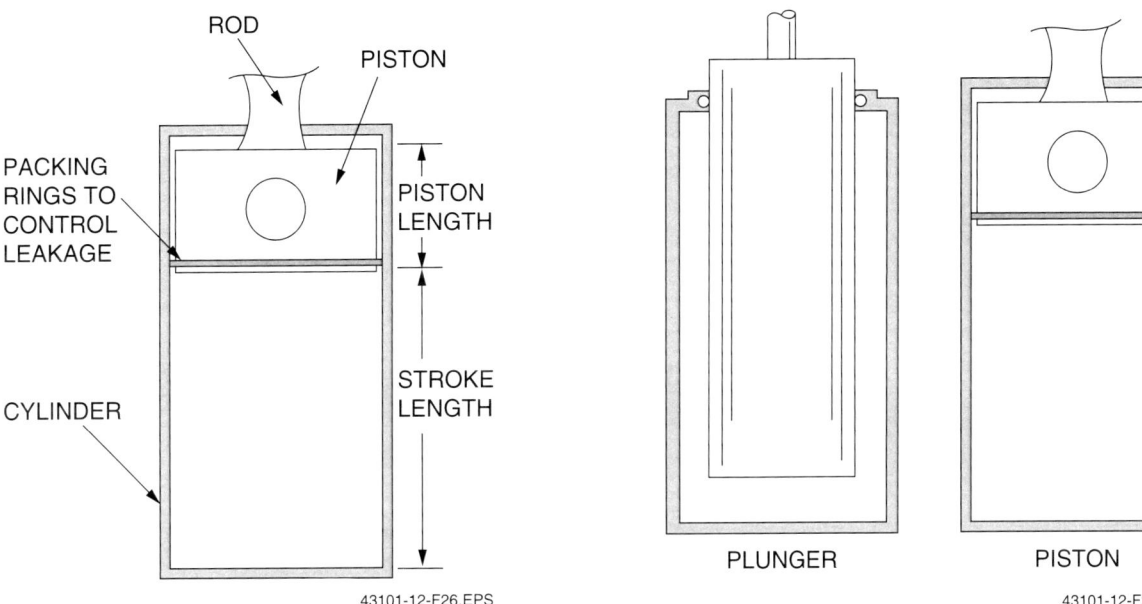

Figure 26 Piston.

Figure 27 Plunger and piston.

sterile conditions. Most systems that operate at extremely low temperatures also use stainless steel valves because they are less brittle at low temperatures than iron or carbon steel. Highly corrosive chemicals require specialized valve bodies that are made of plastic, rubber, ceramics, or special alloys.

5.4.1 Dump Valve

When the hydroblasting technician releases the trigger or steps off the pedal, the water is pre-vented from traveling through the lance while under pressure. Although under less pressure, the water that flows through the lance is still danger-ous, and caution should be used. The dump valve allows the water flow to be immediately diverted from the lance. Dump valves may be mounted on the shotgun (*Figures 29 and 30*) or controlled by a foot pedal. When the hydroblasting technician re-leases the trigger or steps off the pedal, the water is prevented from traveling through the lance to the nozzle. The dump valve must be controlled by the lance operator, or in the case of a flex lance by the

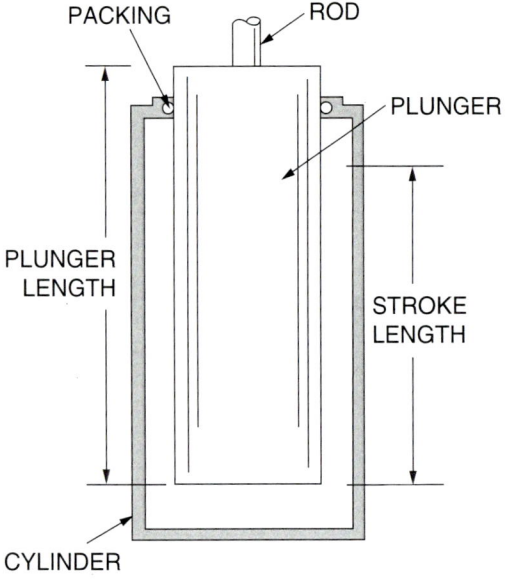

Figure 28 Plunger.

5.4.2 Dry Shutoff Valve

The dry shutoff valve is used to direct flow from the pump to one of the discharge devices. The dry shutoff valve is used when a 100-percent shutoff is required at the lance, or to supply water to two different lances, or to a lance and a line mole.

5.4.3 Relief Valve

All high-pressure pumps are provided with a relief valve, designed to operate when the downstream equipment is not open and the pump is still going. Since the positive-displacement pump produces very high pressures continuously as long as it is running, the relief valve prevents serious damage to the discharge equipment. Portable pumps may have an automatic relief valve or a rupture disk (*Figure 32*) or other emergency overpressure relief on the discharge tank. When the pump is running, the emergency relief system must not be directed into occupied areas. Never defeat the operation of relief valves. If the valve is relieving pressure inappropriately, have the valve

person closest to the mole nozzle. The foot pedal dump valve (*Figure 31*) is used with the flex lance because both hands are required to control and feed the line into tubes. As long as the foot pedal or the trigger is held down, water is directed to the nozzle of the lance; when the pedal is released, the water is diverted to a larger pipe orifice or back to the pump intake and the water flow to the nozzle drops to a safe level. Whether the dump valve is operated by hand or foot, the valve must be protected from accidental operation by some form of surround. The purpose of the foot pedal valve is to activate whip or dump pressure.

When the pump is shut down, the pressure on the discharge line must be released. Otherwise, the discharge line may still retain pressure, which could accidentally be released.

Some dump valves are electrically controlled. Voltage of electrically controlled dump systems should be 24 volts or less.

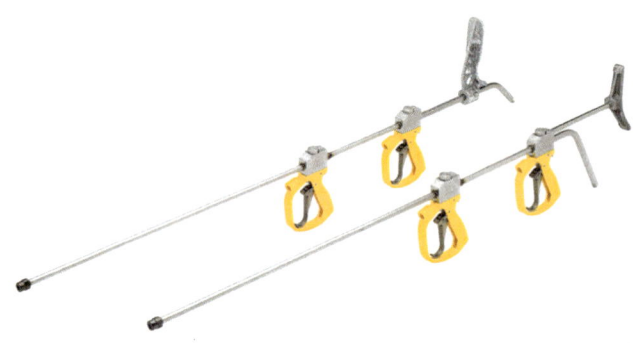

Figure 30 Double-dump shotgun.

Figure 31 Foot pedal dump valve.

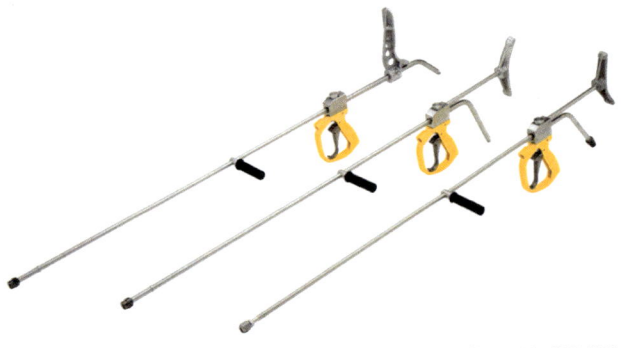

Figure 29 Single-dump shotgun.

checked or replaced. Do not operate the system without a functional relief system. A relief valve should be set to relieve pressure when the pressure reaches 120 percent of intended operating pressure. Always replace a rupture disk with another rupture disk, never with a substitute.

5.4.4 Bypass Valves

Every hydroblaster generates a fixed amount of flow at a given pressure. Bypass valves are a way of diverting some or all of the output of the hydroblaster away from the user. This allows the user to determine the amount of flow and pressure desired, with the balance of the output of the pumps flow going either back to the water tank or to the ground. Were it not for the bypass valve, the user would have maximum pump flow at whatever pressure was selected. The bypass valve starts fully open in bypass mode. Once the hydroblasting technician has activated the foot pedal or shotgun trigger, the operator can dial in the pressure and flow desired independent of the pump's actual output.

5.5.0 Hose and Filter

The water supply to the pump is obtained in several possible ways. The source of the supply and the available connections on the pump determine the connection types used. On the suction side of the pump, filtration is necessary to protect the pump and equipment from damage (*Figure 33*). The hose on the discharge side of the pump will be a very high-pressure hose (*Figure 34*), made of layers of reinforcing materials and high-strength hose walls. The hose should have a burst-rating

minimum of 2½ the intended operating pressure. You must take care of the hose. Scuffs and abrasions to the surface can weaken the hose to a failure point. Make certain that the hose is not laid across a possible traffic area where it might be run over by heavy machines. Also, be conscious of pedestrian traffic patterns around your work area as the hoses can constitute a tripping hazard. Inspect the hose and all connections every shift to be sure there are no damage points or weaknesses to be found. Be sure that all fittings and connections are compatible with the hose and rated at the high pressures used, and test the hose and fittings together. In the case of the dump valve, the pressure rating is usually marked on the valve housing itself (*Figure 35*). All pressure hoses in use must be hydrostatically tested quarterly by bringing the hose up to a test pressure (TP) at either 1½ times or 20 percent above the intended maximum allowable working pressure (MAWP). The longer the hose, the greater the pressure drop will be. Keep your hose as short as possible.

Documentation of the hose tests must be recorded to indicate having been tested at a minimum, quarterly. The hose-test documentation

43101-12_F33.EPS

Figure 33 Water filter.

43101-12_F32.EPS

Figure 32 Rupture disk.

should include the type (e.g., 10,000, 20,000, and 40,000 psi), size, and identifying markings (e.g., serial number or special numbering system) on the hoses to ensure all hoses get tested. Hydroblasting companies must pressure test and band or tag high-pressure hose assemblies quarterly as a minimum. The band color and/or tag must reflect the current quarter in which hoses have been tested. Also, color-coding each piece of equipment makes it easier for workers to be sure that the correct components of each assembly are being used.

5.6.0 Lances

There are several varieties of lances. Each type gives technicians choices for particular operations. The type of lance used for cleaning large areas, called a shotgun, is rigid (for safety), is not less than 66 inches long, has a shoulder brace to absorb the reaction of the high-speed flow, and has either a trigger or a lever to initiate flow (*Figure 36*). The trigger mechanism is now available on some lances as a cartridge, allowing the unit to be replaced in minutes. These shotguns can be used to clean pipe or tubing, but since the lance is not very long in comparison to a 20-foot length of pipe, it is only suitable to clean the first two feet of tube.

The flex lance, rigid lance, or the line mole is used on longer pipe. It is inserted and run all the way through the pipe and then withdrawn. On flex lance, rigid lance, or line mole applications, the hydroblasting technician feeds the lance through the pipe. In the case of a pipe of larger diameter than about twice the diameter of the flex lance, the end of the flex lance is attached to a stinger, to which the nozzle is attached. The stinger is a piece of rigid pipe or tube that is at least 1½ pipe

diameters long, which prevents the nozzle from reversing direction. Rigid lances are available in lengths up to 20 feet; a lance that long frequently requires more than one handler. In such a case, the operator closest to the nozzle is the person controlling the foot pedal. The hydraulic lance delivers a water jet from a hydraulic pump.

Antiwithdrawal devices are designed to prevent the lance from escaping control and injuring personnel. The antiwithdrawal device (*Figure 37*)

Figure 35 Valve pressure rating.

Figure 36 Shotguns.

Figure 34 High-pressure hose.

is used especially when cleaning larger diameter pipe. The frame bolts to the flange, or is otherwise clamped in place, and the lance and stinger are inserted through the clamp in the middle. The clamp is moved down to close on the flex lance (*Figure 38*), and the coupling at the base of the stinger is then unable to exit the pipe. The other antiwithdrawal device is the handheld antiwithdrawal device (*Figure 39*), which is used for smaller diameter pipe. Here, the handle is passed over the stinger and onto the lance. The C-shaped collar piece is slid over the lance, and settled into the socket in the end of the tool. The nut is then slid into place and tightened to secure the collar piece. Again, the coupling between the flex lance and the stinger cannot pass through the collar, but the flex lance can feed through (*Figure 40*).

There have been many recorded occurrences of a nozzle coming out of a pipe, tube or tank opening with the jets going. The nozzle will be mov-

43101-12_F39.EPS

Figure 39 Handheld antiwithdrawal device.

ing very fast and can seriously injure you with its jets. If possible, use a stinger and other antiwithdrawal and shielding equipment as well. In one case, a crew decided not to use a stinger and tried to clear a blockage in a 12-inch vertical tube with a line mole. An O-ring came loose in the nozzle and blocked some of the propelling jets. The nozzle turned 180 degrees and came out of the pipe.

43101-12_F37.EPS

Figure 37 Antiwithdrawal device.

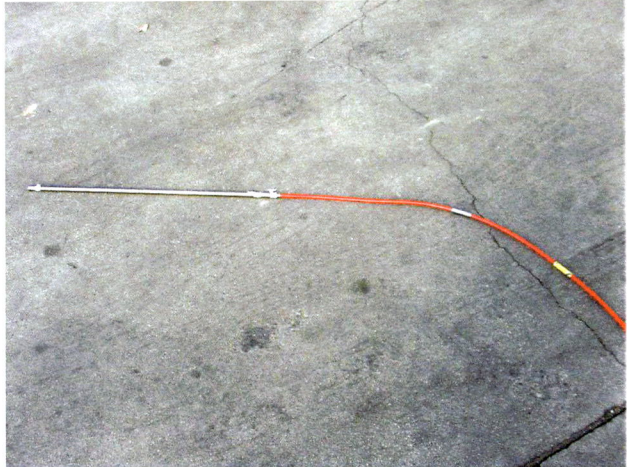

43101-12_F38.EPS

Figure 38 Flex lance.

43101-12_F40.EPS

Figure 40 Using the handheld antiwithdrawal device.

The hydroblasting technician was hospitalized and was unable to work for more than a year.

5.7.0 Nozzles

The nozzle is also called the tip, jet, or orifice of the lance (*Figure 41*). High-pressure water comes into the nozzle, where it is forced through one or more tiny holes. The shape and placement of the hole or holes determine the shape of the spray. The size of the hole determines the velocity of the water jet. Depending on the application, nozzles are either 0-degree straight tip or 15-degree fan jet. A fan tip produces a fan-shaped flow pattern from the nozzle. Materials may be hardened stainless steel, tungsten carbide lined, or sapphire lined. Tips may wear from the constant high-velocity water. If the jet changes shape, or loses effect, shut down the pump and change the tip. The tip is threaded onto a lance or into a holder on the end of the shotgun, with the male threads Teflon® taped just short of the end, so as not to lose any Teflon® into the stream.

The nozzle may be a puller, which has holes drilled at a 45- or 60-degree angle to the rear, or a cutting tip, which only has one hole, facing forward. The cutting tip is most commonly used, and may be found in a number of sizes, based on the internal diameter (ID) of the aperture. The tips are numbered by the size of the aperture, from 0.15 (0.015 inches) to 0.32 (0.032 inches), and then from number 2 (0.036 inches) to number 40 (0.156 inches). Tables are available from manufacturers giving flow rates for different apertures at different pressures.

The other type of tip, used most often for external cleaning, is the rotator or spin nozzle, which

43101-12_F41.EPS

Figure 41 Lance tip.

has the nozzles set at an angle, so that the end spins. This tip is set in a tip housing that turns freely on the lance.

6.0.0 HYDROBLASTING APPLICATIONS

Hydroblasting and water jetting are both names for the same process. In America, *hydroblasting* has come to be the name commonly used for cleaning and surface-preparation operations, while water jetting is more often used to describe machining with high-pressure water, and the use of abrasive mixtures for cutting and shaping, and for coating removal. A third specialty is hydrodemolition, in which ultrahigh-pressure water jets are used to scarify and remove degraded concrete. The technology is especially effective for repairing reinforced concrete, as the water removes all the damaged concrete without damaging the reinforcement bars.

Water jetting is used for cutting softer metals and plastics, using a computer-controlled jet to cut shapes. The jet is a mixture of abrasives and water, and the tip is frequently industrial sapphire or a similar material, to resist wear from the abrasive and water. This technique is very quick, and produces a very accurate cut. The fact that the material is not heated in the process is a major factor in the spread of this technology. Ferrous metals are more likely to be cut with plasma jets or oxyacetylene burn tables, as the metal is generally harder than the materials on which water jets are used.

Hydroblasting is used for cleaning and clearing pipe and pipe assemblies, such as heat exchangers. The technology is very well designed for such uses, and does not damage the pipes. It is used for cleaning the inside of vessels. Chemical cleaning is a surface preparation method using chemical conversion.

6.1.0 Rigid Lancing

A rigid lance is a rigid tube at the end of a hose that supplies water for hydroblasting or a jet of oxygen to a furnace or hot flame for cutting. They are also used to clean heat exchangers. Rigid lances are the best choice for cleaning heat exchanger tubes because of their safety and productivity. They keep the operator farther away from the hazards of high-pressure water as well as chemical spray and exposure. Compared to flexible-lancing systems, rigid-lancing systems produce more forward force for unplugging and more torque for drilling.

6.2.0 Hydroexcavation

In hydroexcavation, also referred to as vacuum excavation, potholing, and hydrodigging, pressurized water or air and a vacuum are used to remove debris in a nonmechanical, nondestructive manner. Compared with conventional digging methods, this process allows for quick, clean, precise debris evacuation that requires less backfill, fewer workers, and less restoration, as well as having less of an environmental impact.

6.3.0 Industrial Cleaning

In hydroblasting, industrial cleaning of heat exchangers and vessel interiors make up the majority of cleaning projects. Internal pipe cleaning is also frequently required. Usually, 10,000 psi is sufficient to clean most tubes; however, the standards for stainless steel tubes at chemical plants are often higher and require that higher-pressure hydroblasting be used.

6.4.0 Sewer Cleaning

Sewers can become clogged by sewage backup caused by landslides, mudflows, floods, and root intrusion. High-pressure hydroblasting quickly and safely removes built-up residue and debris from clogged sewer pipes and also is used for routine preventive maintenance.

6.5.0 Tube Cleaning

Tube systems are cleaned by hydroblasting in many different industries. Sewer blockages are cleared with a line mole (*Figure 42*), without having to excavate or risk harm to the pipes. In industrial applications, pipe bundles are cleaned for many kinds of chemical plants and refineries, and plant turnarounds are expedited by clearing products from pipelines with high-pressure water. Heat exchangers are cleared of scale and debris very quickly and economically with a flex lance or whip, or with a rigid lance. The exterior of heat exchangers is frequently cleaned with a shell cleaner (*Figure 43*). With the rigid lance, in order to clean the whole length, the lance may be as much as 20 feet long, and may require more than one person to handle the length. The operator closest to the nozzle is always the lead operator, and controls the dump valve.

6.6.0 Tank Cleaning

Tanks in the chemical and power generation industries are cleaned by hydroblasting, which releases sludge and scale from the walls, and helps to remove settled material from the bottom of the tank. This is confined-space work, and requires strict adherence to safety rules. Recently, new automated equipment has been developed to allow tanks to be blasted clean without the operator having to enter the tank (*Figure 44*). A telescoping lance with a 3-D nozzle (*Figure 45*) is inserted into the tank, preferably from the top. This allows jets to cover the entire interior of the tank without the material washing out onto the worker. Automation prevents the worker having to enter, and minimizes the risk of injury that is always present in confined-space work. The same rotating nozzle system is used for delayed coker unit cleaning, in oil refinery applications, and in cleaning ships' boilers.

6.7.0 Structural Cleaning and Concrete Repair

Many underwater structures, including lock gates, sewer grates, and other essential functional units, have been infested and compromised by

43101-12_F42.EPS

Figure 42 Line mole.

43101-12_F43.EPS

Figure 43 Shell side cleaner.

Figure 44 Automated equipment.

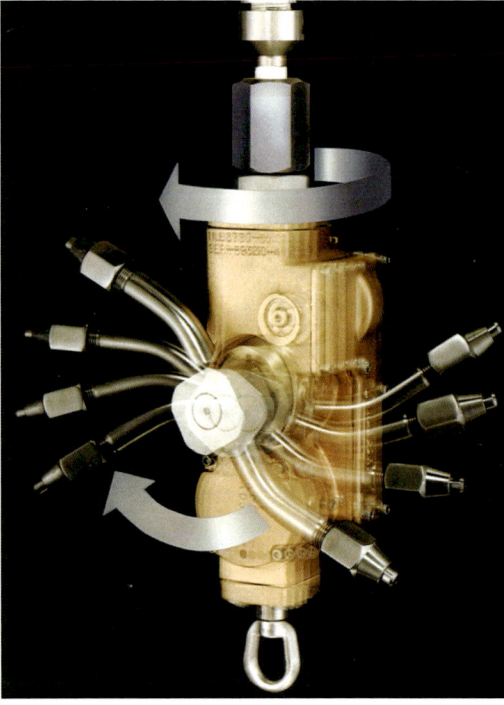

Figure 45 3-D nozzle.

an intrusive shellfish called the zebra mussel. It has been found that hydroblasting removes zebra mussels from their attachments, quickly and relatively easily.

Concrete can be cleaned with ultrahigh-pressure equipment, and only the concrete that has been broken down chemically is removed, leaving a clean and rough surface to which the new concrete can attach. With or without abrasives or other additives, water jetting can clean stains and built-up materials from any sufficiently hard surface.

6.8.0 Cutting

High-pressure water is used to cut metal tanks, towers, and vessels that cannot be cut using a flame (oxyacetylene) due to a possible fire or explosion from the contents that may be explosive or flammable. The high-pressure water can be used to cut metal up to six inches thick. The water is used in combination with an abrasive, such as sand or garnet, at very high velocities to make the cut. The abrasive and water wear the surface from the action until the surface is cut into two pieces. Other uses of high-pressure water for cutting may not use abrasives. Hydroblasting is used to cut concrete on roadways or under demolition. Also, high-pressure water is used to cut precision shapes on a water table for many industries. Again, water is a versatile tool that has many uses in industry.

6.9.0 Grounding Hydroblasting Equipment

Static electricity can be generated by hydroblasting equipment, the surfaces being blasted, and exhaust ventilation systems (fans and ductwork). It can shock employees and cause fires and explosions by igniting flammable or combustible atmospheres or materials.

Hydroblasting equipment shall be grounded when using demineralized water (*Figure 46*). Some plants may require equipment to always be grounded regardless of the type of water used. The hydroblasting technician should wear appropriate gloves and boots for insulation from static electricity. Additionally, blast hoses can be constructed with antistatic rubber linings or fitted with a ground wire or similar mechanism to dissipate static electrical charges.

7.0.0 PERSONNEL

A hydroblasting team is typically composed of three people: a crew leader and two technicians. The technicians may rotate between attending the pump, operating the lance, and maintaining a line of sight. The crew leader, or team leader, is responsible for conducting the pre-job safety meeting and reviewing the job safety analysis (JSA) with the crew. Any personnel involved in hydroblasting operations must be physically fit for the work. The back pressure from the shotgun is a steady drain on the operator, and the vibration, noise, and constant attention requirement

43101-12_F46.EPS

Figure 46 Bonding and grounding a 3-D tool to a tank.

43101-12_F47.EPS

Figure 47 Worker stances at different pressure levels.

makes fatigue a frequent hazard. A moment's carelessness can have permanent consequences. Technicians are taught to take stances designed to prevent balance disruption while controlling the gun (*Figure 47*).

The hydroblasting technician is responsible for whip checks, directs the jet, and controls the dump system. It is in everyone's interest that all safety precautions be followed, and the hydroblasting technician is the one closest to the nozzle. In moling or rigid lancing, the rigid lance and hose are long and may have several operators handling them at the same time; if so, the operator closest to the nozzle has control of the dump system. When the pressure is being brought up to operational levels, the hydroblasting technician must be allowed to feel the back pressure at the same time. The pressure must not be changed without warning the operator. This is because the pressure change also changes the back pressure, and if the operator is not warned, he or she may lose control of the lance or shotgun.

7.1.0 Heat Stress

Workers are at risk for heat stress when outdoor or indoor temperatures are high. Heat stress can cause heatstroke, heat exhaustion, heat cramps, and heat rashes.

Heatstroke is the most serious form of heat stress. Its symptoms are a rapid rise in body temperature, excessive sweating, an inability to cool down, hallucinations, chills, confusion and dizziness, and slurred speech. Heatstroke can cause death or permanent disability if emergency treatment is not received. Therefore, it is imperative for coworkers to tell the supervisor and call 911 as soon as possible. The person should be moved to a cool, shaded area and sprayed, sponged, or showered with cool water.

Heat exhaustion occurs when the body loses an excessive amount of water and salt, usually due to excessive sweating. Employees at greatest risk for heat exhaustion are older, have high blood pressure, and work in a hot environment. Some symptoms of heat exhaustion are heavy sweating, extreme weakness or fatigue, nausea, moist and/or clammy skin, muscle cramps, and fast and shallow breathing. A co-worker with heat exhaustion should be moved to a cool, shaded or air-conditioned area, drink plenty of water, and take a cool shower, bath, or sponge bath.

Heat cramps usually affect workers who sweat a lot during strenuous activity. Their excessive sweating decreases their bodies' salt and water levels. Symptoms are muscle pain or spasms in the abdomen, arms, or legs. Workers with heat cramps should stop all activity and sit down in a cool place, drink plenty of nonalcoholic fluid. They should not return to strenuous work for a few hours after the episode because further activity may lead to more serious symptoms of heat exhaustion or heatstroke.

Heat rash is skin irritation caused by excessive sweating during hot, humid weather. A heat rash resembles a cluster of red pimples or small blisters. It is most likely to occur in areas of the body such as the neck and upper chest, the groin, under the breasts, and in elbow creases. If heat rash occurs, the worker should try to work in a cooler, less humid place if possible and keep the affected area dry.

Employers should schedule jobs in hot areas for cooler months and hot jobs for cooler parts of the day. They should help workers become acclimated to working in hot conditions by gradually increasing the amount of work they do in these environments. They should provide workers with cool drinks, rest periods, and cool areas for work breaks. When temperatures are high, workers should wear no more than two layers of clothing.

Workers in hot environments should avoid exposure to extreme heat, too much sunlight, and high humidity when possible. They should wear light-colored, loose-fitting clothing and be aware that personal protective equipment may increase their risk for heat stress.

7.2.0 Cold Stress

In very cold conditions, workers may be at risk for cold stress. Cold stress may be brought on more or less quickly, depending on how well acclimated individuals are to cold temperatures. Cold stress may begin to occur sooner in warmer climates than in colder parts of the country because workers there are less acclimated to cold conditions. The wind chill factor also plays a role in the development of cold stress.

There are four basic types of cold stress, each of which is discussed in the following paragraphs.

Hypothermia occurs when the body loses heat faster than it can produce it, resulting in an abnormally low body temperature. When a person's body temperature becomes too low, the brain is affected and the person may not be able to think clearly or move well. As a result, the person may not be aware of what is happening and thus may not be able to seek help. Some symptoms to watch for among coworkers are shivering, confusion, disorientation, blue skin, and, potentially, loss of consciousness. If a worker exhibits any of these symptoms, first alert a supervisor and seek medical assistance. The affected individual should be moved to a warm room or area. Offer the person a warm beverage. If the person has no pulse, begin cardiopulmonary resuscitation (CPR).

Frostbite is an injury to the body caused by freezing. Symptoms include reduced blood flow to the extremities, numbness, tingling or stinging, aching, and changes in skin color (*Figure 48*). Workers experiencing frostbite should be brought to a warm room or area. The affected area should be placed in warm water. Do not allow the person to rub the affected area, which can cause more damage.

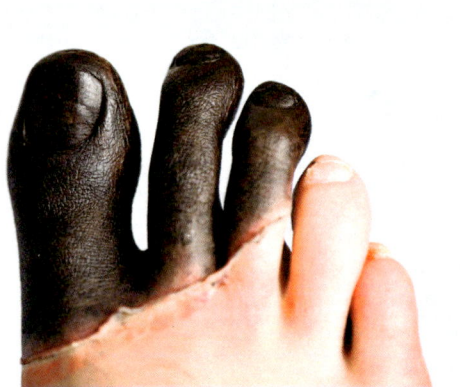

43101-12_F48.EPS

Figure 48 Frostbite.

Trench foot is caused by prolonged exposure to cold and wet conditions. Wet feet lose heat about 25 times faster than dry feet. Symptoms include reddening of the skin, numbness, swelling, blisters, and bleeding under the skin. Workers with trench foot should remove their footwear and wet socks, then dry their feet and avoid walking.

Chilblains are caused by repeated exposure to cold temperatures that results in damage to small blood vessels in the skin. It causes redness and itching, usually on the cheeks, ears, fingers and toes. This damage is permanent. Workers with chilblains should avoid scratching the affected area, slowly warm the skin, use anti-itch cream, and keep the area clean and covered.

Employers should schedule jobs in cold areas for warmer months, as well as later in the day when temperatures are somewhat warmer. Employers should also provide workers with warm drinks and warm areas for break periods.

Workers should avoid exposure to very cold temperatures as much as possible. If working in cold environments cannot be avoided, appropriate clothing should be worn. Wear several layers of loose clothing. Be sure to wear extra covering of the ears, face, hands, and feet if possible.

8.0.0 SETUP

Setting up includes paperwork. Any company for which you work will probably have some sort of pretask checklist. Be sure that you read and understand the purpose of each question.

When setting up equipment for work, it is always wise to take the time to do it right. Look at the area where you will be working. The following issues must be considered in order to do this work safely:

- *Location* – Is the work moveable, and if so, is there a better location for the blasting? It would be best to be out of the major traffic patterns normal for that part of the site, and a safe distance from other work operations. If there is a permanent location set aside for hydroblasting, these issues should have been considered. It is also important that your pump and other equipment be sited out of the possible line of fire of the blasting.
- *Drainage* – Water runs downhill, and you need to be sure that the water is not a problem for anybody else, and that it does not endanger electrical systems or other workers. You are responsible for avoiding problems. If you are to be hydroblasting in any location, you must determine what regulations may pertain to the work. While you are using pure water, what

you are washing off may not be safe or legal to discharge into storm drains. The US Environmental Protection Agency, under the Clean Water Act, has the authority to require a permit from your state if you are discharging wastes into the streams. Municipalities and states have specific criteria for what is permitted.

- *Area isolation* – Isolation and ventilation are the primary means of preventing or minimizing exposure to airborne contaminants during hydroblasting. However, when such controls cannot keep exposures below OSHA-specified levels employees must use National Institute of Safety and Health (NIOSH)-certified respirators appropriate for the types and concentrations of airborne contaminants present during hydroblasting. In all cases, respirators should be donned before entering contaminated work areas and removed only after leaving. Hydroblasting technicians must wear NIOSH-certified Type CE respirators when working in enclosed or confined spaces or using abrasive fluids that contain more than 1 percent crystalline silica. Because of the tremendous power of hydroblasting water jets, you must be certain that no one walks into the range of the lance. You must have barricades and hydroblasting tape (*Figure 49*) around the area beyond the effective range of the blasting equipment. It is not enough to have a simple danger tape; make sure that notices stating that high-pressure water jetting is going on are posted in the area. If you do not have specific water jetting tape, use a red danger tape. In the event that it is not possible to isolate the area, sturdy barriers or panels must be erected to shield operations. Where hoses cross a roadway that cannot be closed to traffic, the hoses must be protected from vehicular damage by a covering ramp. Remember, if someone is injured because you did not warn the person of the danger involved in blasting, you are responsible for his or her

43101-12_F49.EPS

Figure 49 Hydroblasting tape.

injury. If there is more than one lance in operation, side shields must be in place between the operators to prevent spray problems.

Make sure that everyone who might come near the site knows that they must not enter the area while the technician is blasting. Talk to the area supervisors and make certain that they tell other workers in the area to stay outside the perimeter. When the system is under pressure, all personnel shall not be within 6 feet of pressurized connections unless the equipment is shielded or guarded in some manner (including weep holes). The technician is not to be approached while the lance is in operation. The pump operator, in particular, should always pay attention to what is going on around the hydroblasting area, even if it is beyond the barrier. Someone could be injured, or sprayed with water, even if they are outside the safety area, creating an unsafe condition, especially around high-voltage electrical systems. There should always be another worker watching for possible hazards.

Make sure there is no loose material around the area, or loosely attached to the target. The jet could strike a small object and throw it very hard. Debris could also become a tripping hazard. Chock the tires to make sure the pump combination doesn't roll. An emergency shutdown procedure is typically used to protect workers, equipment, and the work environment from hazardous fluid.

- *Water supply* – Check the water supply. It may be potable water, water used for fire fighting, or reclaimed water. But it cannot be sour water, water with a high organic content, as it may obstruct valve and nozzle functions.

8.1.0 Beginning the Shift

For each shift, check the area for any changes that may have occurred. If some new circumstance has arisen that may change the risks, communicate any problems with supervisors and the affected personnel. Every shift, before commencing operations, perform the following checks:

- If there is frozen water in the system, flush hoses before installing the tip. If you are working in freezing conditions, follow manufacturer's recommendations for antifreeze protection. Do not run the pump to pressure until water runs freely from the inlet to the outlet.
- Check the tires, brakes, hitch, lights, safety chains, and mechanical condition of the trailer, if you are using a trailer-mounted unit.

- Whenever you use tools on equipment, use the right tools. Don't try to tighten or loosen couplings with pliers. Use the correct size of wrench.
- Check every inch of hose for visible flaws, cracking, or scuffing. If the rubber coating is worn, and you can see the fibers of the wrapping, make sure the fibers are not broken. Make certain that the lance and the hose are rated for a burst pressure of at least 2½ times the maximum expected pressure. The manufacturer's markings should give the maximum permissible pressure. Be sure to check all connections, making certain that all of them are firmly seated. Check for damaged threads on all connections. Check the quick-connect couplings to make sure the sleeves are all the way closed, and pull on the hose to be certain the coupling is firm. Be sure that the antiwhip cables (whip checks) are in place on every quick-connect coupling in case the coupling comes apart under pressure. If a coupling were to come loose without a whip check, it might lash about from the recoil of the water, and the coupling might injure someone. Another protective measure for technicians is a safety shroud, which covers the hose fitting at the place where it is connected to the gun. Make sure the hose is not laid over sharp edges; when the pump is running, the hose will vibrate, and could be damaged. Make certain, also, that the hose is laid out and stored without kinks, as a kink can damage the integrity of the hose. *Figures 50–52* show various types of clips and cables.
- Inspect and clean the dump valve every day. It has to work properly every time you operate the lance. The dump valve, whether in a foot pedal or a shotgun trigger, is a spring-loaded,

43101-12_F50.EPS

Figure 50 Whip clip.

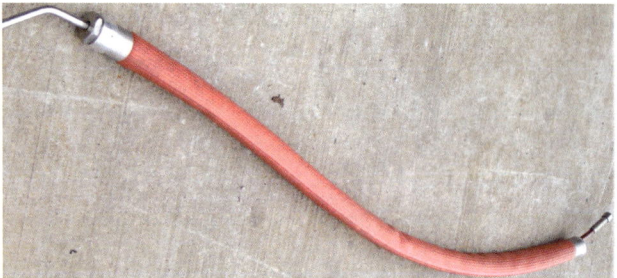

43101-12_F51.EPS

Figure 51 Metal whip clip.

Figure 52 Nylon whip clip.

Figure 53 Water flowing through the lance.

fail-open needle valve. The pedal or trigger compresses the spring, and directs the water through the lance (*Figure 53*). When the pedal or trigger is released, the valve opens, diverting the water to the dump nozzle (*Figure 54*). If the dump valve does not work, the pump must be shut down, the pressure released, and the valve repaired or replaced. The cartridge-type valve used is accessed by unscrewing the cartridge from the valve casing. On the foot pedal, loosen the valve by putting a 10-inch adjustable wrench on the nut between the quick-release coupling and the valve (*Figure 55*), and unscrew the cartridge and coupling until the valve stem can be pulled free. On the shotgun, put the wrench on the nut on top of the trigger assembly housing, and unscrew the cartridge and remove it from the case (*Figure 56*).

- Check the pump engine. Check the lubricant level, coolant, hydraulic fluid, and fuel levels. Make sure the engine is adequately fueled. Be sure before you start of the correct fuel for the driver, and do not allow anyone to smoke around the fueling operation.
- Inspect the pump. Check the pump lubricant and gearbox lubricant fluid levels.
- If there is a powered hose reel, check lubrication and hydraulic fluid levels, if applicable.
- Check the condition and positioning of all the shields. If you are cleaning tubes or tube assemblies, place deflectors or shields on the opposite end of the tubes to keep the water from coming out too far.

Figure 54 Water being dumped.

Figure 55 Removing the foot pedal dump valve.

Figure 56 Removing the shotgun dump valve.

- Check the water filters. Water, even clean tap water, often contains suspended particles, and the particles could clog the shutoff valve or an aperture at the nozzle, either of which could become a hazard. Water must be filtered before entering the system, as the size of the opening at the nozzle could produce a stoppage by relatively small particles that could be harmful to equipment or possibly dangerous to the workers. Check the tip for clogging; if it is fully or partially plugged, stop working, shut off the pump, and use the cleaner to clear the tip. Valve seats and pump pistons are already subject to wear from the pressure of the water; abrasion from particulate matter makes the wear worse. This is especially hazardous in a multiple nozzle operation because the recoil could become unbalanced suddenly, causing the lance to jerk one way or another. In the case of a line mole application, in particular, without a stinger or other protection, if some or all of the back jets are blocked, the nozzle can either blast out of the pipe backwards or turn around and come out so as to blast the operator. Make certain the pump is turned off and inspect and replace or clean the filter as follows:
  - Turn the wingless eyes counterclockwise to loosen them. You may have to use a screwdriver or metal rod to turn them (*Figure 57*).
  - Swing the lid out of the way and pull out the filter (*Figure 58*). Wash it and inspect it for damage. If it has holes, replace it.
  - Return the filter, and replace and tighten the lid. Be sure it is sealed properly. Check the tip for obstructions. The best way to check the tip is to run a welding tip cleaner through it, so that the inspection and the repair can take place at the same time.
- If you will be lancing tubes or pipe with a flex lance, make sure the stinger is in place, and mark the lance at least 24 inches back from the stinger, so that the nozzle can't come out of the tube. If you are lancing a tube, and the tube turns out to have an obstruction, the lance must be pulled back and the jet allowed to clear the obstacle. The back pressure from the obstruction can, under some circumstances, push the lance back out of the tube, so do not use a flex lance without a stinger and a warning mark to alert the technician of the length remaining in the tube. It is strongly suggested that an anti-withdrawal frame be used to prevent the nozzle coming clear of the tube or pipe. If you are moling a tube, and the lance begins backing up, dump pressure immediately.

- Inspect and clean and test the trigger and, if you have one, the foot valve. Make sure that the dump mechanisms work freely, with the nozzle out, while flushing the system initially.

Figure 57 Opening the filter case.

Figure 58 Removing the filter.

- If the pump driver is electrical, inspect all connections, junction boxes, switch housings, and power lines. Look for damage, frayed insulation, or loose wires. Be sure that the electrical parts cannot be struck accidentally by water.
- Check all the engine controls, throttle lines, and cables daily.
- If you are blasting aboveground, you must use OSHA-approved scaffolds or some safe form of personnel lift. The best place to rig the hoses is to hang them from underneath the top handrail, wrap them over the handrails, and attach them so as not to pull toward the ground. Substituting with jury-rigged supports is not an option, and the back pressure from hydroblasting is too powerful to work from a ladder. Don't cut corners; hydroblasting jets are capable of cutting through flesh and bone.
- If you find a leak in any connection, shut down the pump and repair or replace it. As in submarines, high-pressure water equipment has no tolerance for carelessness.

8.2.0 Hooking Up the Lance

Follow these steps to hook up the lance:

Step 1 Connect the pressure hose to the pump (*Figure 59*). Push the coupling closed and give the hose a tug to be sure it is firmly closed. When pressure hoses are connected with quick-connect couplings, they must also be provided with anti-whip cables. These cables are placed so as to prevent the hoses from whipping around in case of accidental release. There is a great deal of recoil from water being discharged at pressure from the hose. The whip checks hold the parts of the hose together, and prevent the coupling and hose from whipping around and possibly striking and injuring personnel.

Step 2 Make sure that the pressure control (*Figure 60*) is as low as possible.

Step 3 Having connected to the pump, connect the other end of the high-pressure hose to the dump valve (*Figure 61*).

Step 4 Connect the lance to the foot pedal quick-connect. The flex lance must be marked visibly at least 24 inches from the tip, and again 36 inches from the tip (*Figure 62*). This serves to warn the operator when the tip is approaching the opening, so he or she can release the pressure to dump.

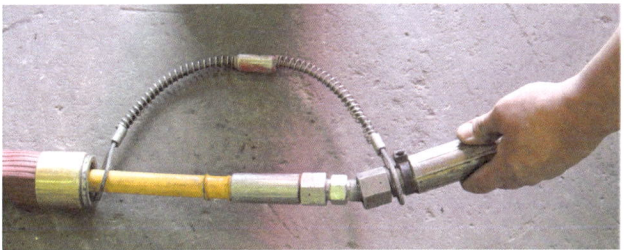

43101-12-F59.EPS

Figure 59 Hooking up the hose.

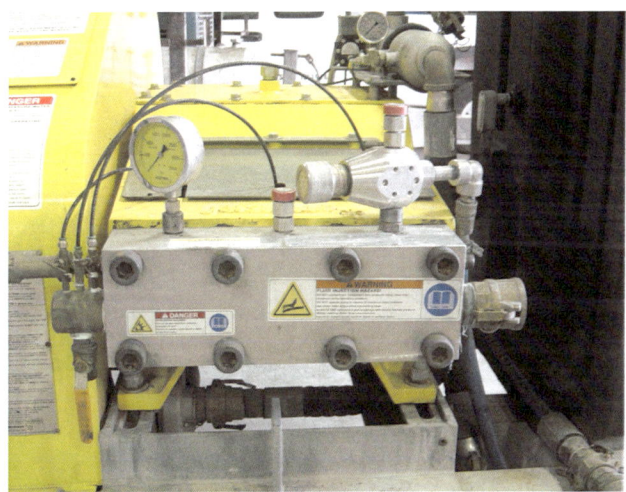

43101-12-F60.EPS

Figure 60 Pump pressure controls.

Figure 61 Dump valve connection.

Figure 62 Flex lance.

The diesel pump is started by the following steps:

Step 1 Turn the starter (*Figure 63*) counterclockwise for a few seconds, to warm up the glow plugs.

Step 2 Turn the starter clockwise to start. Allow the driver to warm up, and when it is running well, bring the engine speed up, by adjusting the throttle (*Figure 64*), to working speed.

When the hydroblasting technician tells you to give the lance pressure, push the pump-engagement lever (see *Figure 64*) to the left to engage the driver with the pump. Start with the pressure at minimum, and slowly bring the pressure up with the pressure-control knob (see *Figure 60*) until the

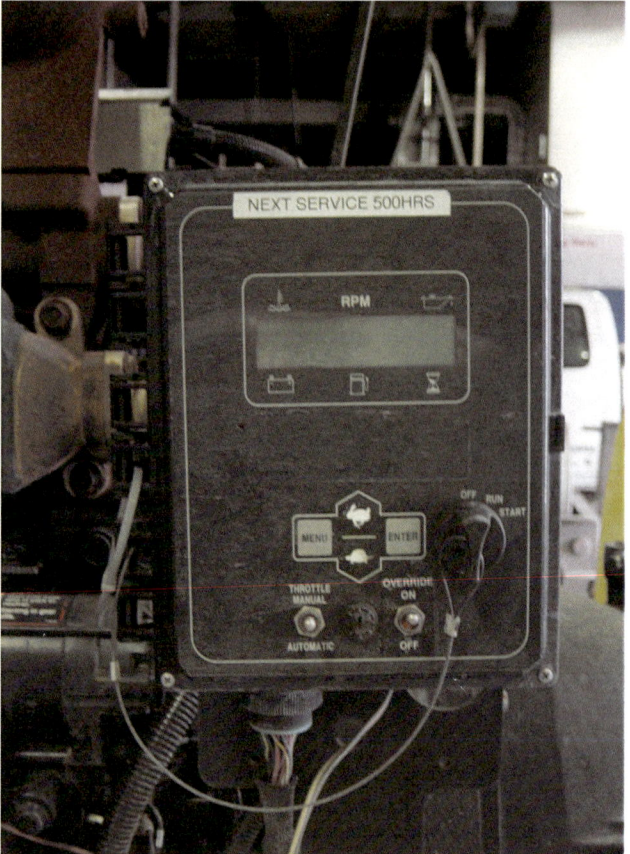

Figure 63 Driver start panel.

hydroblasting technician signals enough or the intended pressure level is reached. If it becomes necessary to cut the pressure to the gun, the pump operator reduces the pressure and then, if so desired, cuts the ignition for the pump.

When the pressure is still at minimum, the hydroblasting technician tests the dump valve. This is done by squeezing and releasing the trigger, or on a foot pedal, by depressing and releasing the pedal. If the valve does not work, stop everything, turn off the pump, and repair or replace the valve. Do not service or adjust any components when the pressure is on. Always shut down first. If the hydroblasting pump begins to make a knocking sound, shut it down immediately and contact your supervisor.

9.0.0 OPERATIONS

When you have completed setup and testing, you are in the operations stage. Confirm with your operator the proper hand signals prior to performing hydroblasting. It is not in the scope of this text to instruct you on the specific operation of the equipment, but a few general observations may be useful to begin. When you have

Figure 64 Pump and driver controls.

completed the hands-on portion of your training, you will be prepared to operate as a hydroblasting technician.

Before bringing the pump up to high pressure for cleaning, start at a low pressure, without a nozzle on the lance. Flush out the system to remove air bubbles and particles, and check the operation of the dump system. Look for leaks anywhere on the system. If any component is malfunctioning or leaking, do not attempt to repair the connection under pressure; shut down the pump, and find out why. Do not ignore leaks or other problems and go on to high-pressure work; catastrophic failure of components could result, with possibly serious consequences. If the hose begins to pulsate when running, the pump may not be receiving enough water at the suction side. Turn it off and check for problems. Another indication of insufficient water supply might be a banging sound from the pump. This is called cavitation.

When you have flushed the system, turn it off and install the nozzles. Now test the system at low pressure, gradually bringing the system up to working pressure. If there are any leaks, shut the equipment down and repair them.

> **WARNING!**
>
> Never try to repair or tighten connections under pressure.

The operator will notice a distinct jolt when the dump valve or dry shutoff valve is operated. This is called a line pulse, and should be prepared for. There are limiting devices available for this occurrence. One such device is a chamber containing air attached to the line, which uses the compression of the air to absorb the line-pulse impact.

9.1.0 Shotgunning

The most common hydroblasting technique is shotgunning. When using a shotgun to clean large surface areas, you must be continually conscious of where the tip is pointed. That said, effective cleaning is conditioned by how close the tip is to the surface being cleaned. With a fan jet, the wider the area being covered in a sweep, the less force is available. You will also find that the closer the tip is to the surface, the greater the backsplash.

Be sure to accustom yourself to the back pressure gradually, as you could lose control of the gun, or even fall down, if you are suddenly exposed to the kick. Remember to check your footing constantly, as the combination of water and oil can produce a very slippery surface, and it may not be slippery everywhere. If you feel that you may lose control, go to dump immediately.

9.2.0 Pipe

While the first 24 inches of a pipe should be cleaned with a shotgun, pipes are commonly cleaned now, especially in larger diameter pipes or contexts where the pipe cannot be brought to a site, with a flexibly hosed line mole. The line mole has a jet in the front to cut obstacles, and several smaller jets in the back. These wash the waste back out of the pipe at the same time they propel the mole forward. The nozzle is started into the pipe with the pressure off, and when the water is started, pushes itself through. At the operator's end of the pipe, if it is possible, the antiwithdrawal device holds the hose to keep it centered at the entrance. The reaction from the jets keeps the nozzle centered inside the pipe. The line mole must be controlled very carefully. Do not allow the nozzle to come clear of the end of the tube it is cleaning while flow is still going through the nozzle. The hose must be clearly marked no less than 24 inches from the nozzle, to prevent the nozzle from being accidentally pulled clear of the tube. Make sure you know what has been in the pipe before you start. Even diluted, some chemicals can be very dangerous.

9.2.1 Flex Lance

The flex lance is used for shorter lengths of tubing, such as heat exchanger assemblies. The lance must be marked visibly 24 inches from the stinger, if one is in place, or at least 24 inches back from the nozzle, so that the operator can release the flow before the nozzle clears the pipe. It is possible for the flex lance to turn around in a large enough space so that it is blasting backwards at the operator. In large pipes, always install a stinger between the end of the flex lance and the nozzle. The length of the stinger must be at least 1½ times the diameter of the pipe in which it is used.

In either case, line moling or flex lance operations, make sure there is adequate clearance for water and debris to flow around the head. This prevents the pressure from making a projectile out of the nozzle, or from pushing the nozzle the wrong way.

When the flex lance is used in a pipe more than twice the diameter of the flex lance, the operator rotates the lance by a circular movement of his or her arm, called whipping. This causes the nozzle to rotate inside the pipe and ensures complete cleaning.

9.3.0 Tanks

Hydroblasting tanks, boilers, cooling towers, and other enclosures require particular techniques. The use of a telescoping lance and a rotating T-head with two nozzles allows the operator to work inside without physically being inside. It is still very critical that the nozzles be kept from coming out of the opening while under pressure. You must mark the lance at least 24 inches from the head so that you will not accidentally pull the head clear of the tank. It is best to lower the nozzles into the tank from above to prevent the wash material from interfering with control. In one case, a chunk of material being cut out of a tank struck the lance, and knocked the lance out of the opening, injuring two workers severely.

9.4.0 Cleanup

You will already know before you start what cleanup you are required to do on the work site. This is also part of your work. It is important that you meet or exceed the requirements of the company you are working for, to keep your employer, and the companies your employer sends you to work for, convinced of your value. Always do your job so well that anyone watching you, or seeing your work, would like to employ you. Coil and store all hoses and check them in the process. Check the valves and fittings, and do a careful check on the pump and driver so that there will be no delays in using them on the next job.

10.0.0 Variances

This text is a compilation of standard practices, most of which are common to the industry. You are required to learn, and to practice, the required practices of the company for which you work. Any change to a standard procedure, such as shortening a shotgun extension to allow work inside a tank, or allowing some special system to be used, must be approved by a supervisor, and requires a variance to be issued giving the description of the change and the reason for the variance. Different companies may have different assignments of responsibility for variance. Be sure that you know and follow the requirements established by your company.

The variance form shown in *Figure 65* is to be filled out and signed by one or more of the following: a supervisor, the company safety representative, the technician on site, or the safety officer of the customer. The required information includes a statement of the procedure to be changed, such as a shotgun to be used with a barrel shorter than 48 inches.

The reason for the change in procedure is then filled out; in the above case, it might be because the operator would be working in a confined space, in which the longer barrel was unusable. Next, the advantage in safety must be stated; for example, the operator will be able to maneuver the barrel safely in the space. Finally, the scope of the job is stated; that is, the range of work to be done in this particular case.

No change in standard operating procedure should occur without a signed variance from the appropriate persons. Until the variance has been filled out and signed by the appropriate person or persons, you must not begin work that involves a change from standard operating procedures. Operating procedures are those which have been shown to be safest over most circumstances. Many jobs using hydroblasting fall under the designation of HAZWOPER operations, and require specific procedures to be followed. If you are sent on such jobs, you are required to complete specific HAZWOPER training, and to maintain certification by yearly refresher training.

Deviations or variances from plant or hydroblasting company policy are acceptable only after all other measures have been exhausted. The variances require upper management approval as well as the approval of the plant or owner representative. Examples of these deviations are the use of shotguns less than 66 inches in length and not being able to make use of an attached anti-withdrawal device for flex lancing, rigid lancing, or line moling.

ANCON MARINE
Safety Policy Variance Form

Customer: Unit/Area:
Address: Date:
                                             Shift:
                                             Permit #:
Supervisor: Leadman:

1.   **SOP/Safety Policy being deviated from:**

2.   **Reason for Variance:**

3.   **Equivalent or greater safety obtained by the following**

4.   **Job Scope:**

| Safety: |
| Operator: |
| Ancon Safety: |
| Ancon Supervisor: |

43101-12 F65.EPS

Figure 65 Sample variance.

11.0.0 ULTRAHIGH-PRESSURE HYDROBLASTING

High-pressure water jetting delivers an average of 15,000 psi of pressure with approximate water volumes of 12 gallons per minute and has long been a reliable, and often the only, method of industrial cleaning. Increased pressure with ultrahigh-pressure water jetting is an even more powerful tool (*Figure 66*). Ultrahigh pressure is defined as 20,000 psi and higher. Pressure this high is used to cut away coatings, including rubber tank linings, and ships' hull paint, as well as to remove hazardous chemical waste, fused polymer, to unblock concrete from drains, and to cut heavy duty industrial equipment that is ready for disposal. The special tips used at these pressures may include industrial sapphire tips. The equipment used at these pressure levels must be rated at the higher level. Standard hydroblasting and ultrahigh-pressure hydroblasting equipment must be kept separate.

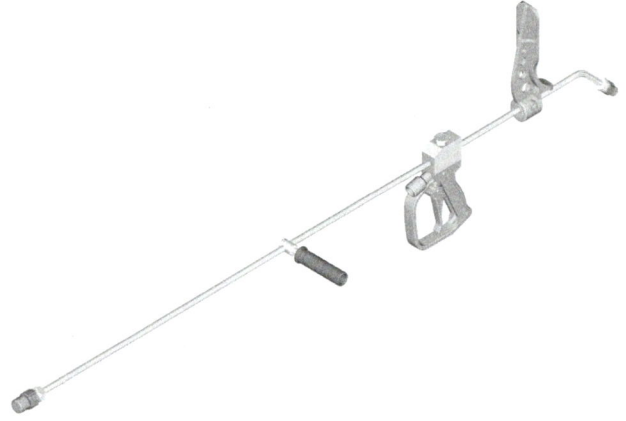

43101-12_F66.EPS

Figure 66 Ultrahigh-pressure hydroblasting.

SUMMARY

Hydroblasting is a highly skilled industrial field, requiring the most extreme diligence and care. If each step is carried through properly, everyone involved can be safe. Examine all equipment carefully before you begin. Follow all applicable procedures in the setup to be sure nothing has been overlooked.

Make sure that all the equipment is compatible and correct for the anticipated pressures. Be sure to use all appropriate safety equipment. Bring pressure up gradually and use the equipment carefully. Leave the equipment in good condition and be sure the site is left clean.

1. The range for pressure used in hydroblasting is from _____ .
   a. 100 to 2,000 psi
   b. 1,000 to 4,000 psi
   c. 1,000 to 10,000 psi
   d. 1,000 to 48,000 psi

2. Impact is the function of mass and _____.
   a. velocity
   b. pressure
   c. volume
   d. flow rate

3. It is necessary to go to a doctor for a small surface wound from a hydroblasting lance if it *is not* bleeding.
   a. True
   b. False

4. Hydroblasting workers need hearing protection when sound exceeds 85 decibels.
   a. True
   b. False

5. If you are cleaning a tank (from the outside), and you have been told that it is a permit-required confined space, and you drop a nozzle into the tank, you should _____.
   a. climb inside and get it, it'll just take a minute
   b. tell your other operator, then climb in and get it
   c. contact the supervisors at the worksite, and arrange a confined space permit and an entry
   d. try to reach it by sticking your upper body inside

6. Which of the following is used on a water blast shotgun?
   a. Water jet
   b. Lear jet
   c. Plasma jet
   d. Fan jet

7. The type of pump used to produce high-pressure water for hydroblasting is a _____.
   a. positive-displacement pump
   b. centrifugal pump
   c. peristaltic pump
   d. rotary gear pump

8. The dump valve commonly used with a flex lance is a(n) _____.
   a. air relief
   b. hand
   c. foot pedal
   d. knife gate

9. The discharge-side hose should be burst rated at minimum _____.
   a. exactly the expected pressure
   b. one and one-half times the expected pressure
   c. twice the expected pressure
   d. two and one-half times the expected pressure

10. The rigid piece of pipe at the connection of a flex lance to a nozzle is called a(n) _____.
    a. shotgun
    b. stinger
    c. antiwithdrawal device
    d. mouse

11. Fan jet nozzles produce a _____.
    a. 0-degree spray
    b. 15-degree spray
    c. 45-degree spray
    d. 90-degree spray

12. The person in control of the dump system is the one _____.
    a. at the pump controls
    b. closest to the nozzle
    c. at the truck
    d. farthest from the jet

13. A line mole is used to clean _____ .
    a. pipes
    b. ship hulls
    c. boilers
    d. tanks

14. Tanks should be cleaned, if possible, from the _____.
    a. top
    b. bottom
    c. side
    d. rear

15. If you notice a coworker suffering from heat-stroke, you should first _____.
    a. call 911 and notify a supervisor
    b. give them water
    c. move them to a less humid area
    d. ask them to remove their footwear and wet socks

16. When frostbite occurs, the affected skin should be _____.
    a. placed in cold water
    b. placed in warm water
    c. massaged softly
    d. treated with isopropyl alcohol

17. If there is ice in the system, *don't* run the pump up to pressure until _____.
    a. the pipe bursts
    b. water stops running through the system
    c. water runs freely from inlet to outlet
    d. the pump warms up

18. When using a flex lance, mark the lance at least _____ from the back of the stinger.
    a. 2 inches
    b. 12 inches
    c. 24 inches
    d. 48 inches

19. If you find a leak while running the gun, you should _____.
    a. ignore it
    b. keep an eye on it, but keep running
    c. shut down immediately and repair it
    d. work around it

20. Before beginning operations, start the pump at low pressure with the _____.
    a. nozzle removed
    b. nozzle in place
    c. shutoff valve open
    d. lance removed

CONVERSION OF ENGLISH MEASUREMENT UNITS TO THE METRIC SYSTEM

English units	Metric units
1 in.	25.4 millimeters
1 ft.	0.305 meters
1 sq. ft.	0.093 sq. meters
1 sq. ft./minute	0.093 sq. meters/minute
1 cubic ft.	0.028 cubic meters
1 lb.	0.453 kilograms
1 gallon	3.78 liters
1 psi	6.89 kilopascals
Degrees Fahrenheit	Degrees Centigrade = (Degrees Farenheit $\times$ 1.8) + 32

Appendix B

GREATER BATON ROUGE INDUSTRY ALLIANCE (GBRIA) SAFETY FUNDAMENTALS FOR HYDROBLASTING

The Safety Fundamentals for Hydroblasting is a set of guidelines developed by a joint committee of hydroblasting contractors and owner or plant-site members of GBRIA in an effort to reduce incidents, standardize plant and contractor requirements, and improve communication between plants and contractors. These standardized guidelines will help raise awareness of hazards and ways to mitigate those hazards among both plant and hydroblasting company personnel.

The following are the minimum guidelines.

1. Hydroblasting companies must:

    i. Provide their employees the computer-based training module on hydroblasting provided at the Safety Council of the Louisiana Capital Area as a prerequisite for employees to gain plant entry.
    ii. Have a documented training prgram that at minimum includes, but is not limited to, energy isolation of equipment; confined-space entry; correct body positioning for tasks; use of shotguns, flex lances, rigid lances, and line moles; and use and installation of antiwithdrawal safety devices.
    iii. Ensure that all employees know how to properly shut down a pump and disconnect its energy source to perform adjustment or maintenance of the cleaning system.
    iv. Have a documented heat-stress policy, provide training on the policy, and establish a work-rest schedule for all employees.
    v. Have a documented personal-accountability policy on at-risk behaviors that will not be tolerated. The policy must be communicated to all employees, including that disciplinary actions can lead up to and include termination of employment. Hydroblasting companies are subject to having their hydroblasting policies and training programs audited for plant-entry compliance with these minimum safety standards.

2. A prejob checklist/job safety analysis (JSA) must be utilized to help identify all job hazards and ensure the safe execution of the job by the crew. This documentation must address, at minimum, the hydroblasting plant-entry policies found in this document. The prejob checklist/JSA document shall be initiated and reviewed with the crew again any time there is a change in job scope, an unsafe condition arises, additional personnel are added to the job, or at each and every personnel shift change.

3. All required personal protective equipment (PPE) must be worn on the job per hydroblasting company and plant requirements, including, but not limited to, the following:

    i. Hard hat.
    ii. Hand protection.
    iii. Appropriate hearing protection must be worn up to and including double-hearing protection.
    iv. Safety glasses with attached side shields. Dark, tinted safety glasses may not be worn during vessel entry or at night.
    v. Monogoggles and/or a face shield must be worn as required, including when the pH of the chemical exposure is low on the acidic side or high on the alkaline side.
    vi. A face shield must be worn at all times as required by any member of the crew where there is a potential for exposure to chemicals and flying debris.
    vii. An approved protective chemical suit that minimizes the potential exposure to harmful chemicals must be worn at all times when hydroblasting.
    viii. Flame-resistant clothing may be required when hydroblasting.
    ix. Safety-toed, metatarsal protective, chemical-resistant boots must be worn when hydroblasting.
    x. An approved escape respirator must be worn by hydroblasting technicians at all times if required.

xi. Special chemical exposure badges must be worn near the facial area and on the outermost garment at all times if required.

xii. Hydroblasting may require the use of an air-purifying or air-supplied respirator. Reduction in the PPE requirements can be achieved when utilizing certain automated equipment and the PPE reduction meets the approval of the plant.

4. Deviations or variances from plant or hydroblasting company policy are acceptable only after all other measures have been exhausted and require upper-management approval and the approval of the plant and/or owner representative. Examples of these are the use of shotguns less than 66 inches in length and not being able to make use of an attached anti-withdrawal device for flex lancing, rigid lancing, or line moling.

5. Special precautions must be initiated before hydroblasting heat exchangers with metal plugged tubes and tubes of questionable metal integrity. If there is pressure on the shell side of the exchanger, approval from the owner must be obtained before proceeding to ensure that the shell-side pressure can in NO WAY affect personnel, even if a tube is compromised during hydroblasting.

6. During and upon the completion of the job, it is the responsibility of the crew to maintain good housekeeping practices on the job site. This includes, but is not limited to, the elimination of slip, trip, and fall hazards, including proper hose placement as well as proper disposal of trash, contaminated PPE, and chemical and/or product waste generated by the cleaning service.

7. All environmental considerations must be given to protecting the plant's work site from chemical leaks, either from the cleaning operations or from fluids in the equipment being used on the job. The hydroblasting company must comply with their own and all plant environmental policies and provide secondary containment and/or use of spill control kits if required by the plant.

8. Barricading of the pumps and job-site hazards is required at every job site and should be placed from a minimum of 10 feet to a maximum of 25 feet from equipment or hazardous environments, or per plant policy. All

barricades should have four sides and be as square as is practical. The barricade that is used should be demarcated by red tape or indicated by the word "DANGER" on tape with a red background, along with a tag stating that hydroblasting work is being done, unless the plant requires a different color-coded barricade.

9. All high-pressure pumps provided by hydroblasting companies must be equipped with a rupture disk assembly rated at a minimum of 1.2 times the maximum allowable working pressure of the pump. An inspection of the assembly prior to use must show visual evidence (color-coded banding or tagging) of at least quarterly inspection of the integrity of the rupture disk.

10. Grounding of high-pressure pumps provided by hydroblasting companies is required where applicable, as per plant and hydroblasting company policy.

11. At least one tire on a high-pressure pump trailer axle must be double chocked against accidental movement, when parked on location.

12. A trained pump attendant is required to be in attendance of pumps at all times. The attendant must be close enough to shut the pumps down in an emergency. The attendant must not leave his/her responsibility for any length of time without another trained attendant relieving that person or shutting the pumps down. This applies in the event of an emergency as well. A pump attendant is not allowed to attend more than two pumps in operation within the same barricaded area.

13. All jobs shall include a direct line-of-sight between the pump operator and the hose end equipment operator/attendant, in case of an emergency. If direct line-of-sight cannot be achieved, radio communication or use of an additional person for line-of-sight must be utilized to assist in quickly de-energizing the pump.

14. All high-pressure hoses should be routed and protected in a manner that prevents vehicular damage and personnel exposure to the hoses. When possible, roadways should be barricaded to vehicular and pedestrian traffic where high-pressure hoses cross roadways.

15. All high-pressure and ultrahigh-pressure hoses with outer rubber or plastic covering construction and assemblies, including shotgun whip hoses and assemblies, must be tested at a minimum up to the working pressure of those hoses. Documentation of the hose tests must be recorded to indicate that they have been tested quarterly at minimum. The hose test documentation should include the type (e.g., 10,000, 20,000, and 40,000), size, and identifying markings (e.g., serial number or special numbering system) on the hoses to ensure that all hoses are tested.

16. Hydroblasting companies must pressure-test and band or tag high-pressure hose assemblies quarterly at minimum. The band color or tag must reflect the current quarter in which the hoses have been tested.

17. Prior-to-use inspection of high-pressure hoses and flex lances must include visual inspection for the integrity of the compression fittings and the outer coverings, as well as no evidence of broken wire braids. No hose or flex lance with broken wire braids may be manually (handheld) used by the employee, e.g., in line moling and flex lancing.

18. If one or more broken wire braids exists when making use of automated line moling or flex lancing equipment in which the operator will not have to manually hold on to the pressurized hose or flex lance, the hose must be taken out of service and discarded from use.

19. All rigid lance assemblies shall include the use of a minimum 3-foot shroud to protect the employee handling the end of the lance. The construction of the shroud (e.g., double SS-braided) shall be sufficient enough to help ensure the safety of the employee against serious injury from the potential for a high-pressure hose rupture. The hose shroud must be inspected for inner and outer integrity (e.g., excessive wear and breakdown of reinforcement wire braiding) on a quarterly basis at minimum.

20. All dump valves must be guarded against accidental activation, e.g., by the foot pedals.

21. The worker closest to the nozzle must be in control of the dump valve, e.g., the shotgunner, flex lance, rigid lance, or line mole operator.

22. Antiwithdrawal devices must be used on all flex lance, rigid lance, and line moling jobs. The antiwithdrawal devices must be attached to the equipment being cleaned. The lance and/or hose must be secured (tied off with rope or duct tape, etc.) for the maximum length of travel while flex lancing or line moling.

23. Antiwithdrawal devices used on tube sheets when flex lancing or rigid lancing must include a snorkel that does not exceed 1-inch clearance from the tube sheet. In vertical tube sheet applications, the maximum gap clearance of the snorkel and the tube sheet must be additionally protected against the potential of the lance to come out under hydraulic pressure. To prevent the potential of a water cut to the foot, that gap (between snorkel and tube sheet) must be protected from the lance operator by utilizing some type of physical barrier, e.g., a 4-inch × 4-inch × 12-inch piece of angle iron or a 2-inch × 4-inch × 12-inch board. There are some situations where manufactured fixed antiwithdrawal devices do not fit for all applications. In these cases, handheld antiwithdrawal devices are used but require a deviation that must be approved by the owner and hydroblasting company.

24. In addition to use of antiwithdrawal devices when line moling, a stinger must be attached to the end of the nozzle, which is at least 1½ times the I.D. of the piping being cleaned. The rigidity of the compression fitting and nozzle length can be included in the overall stinger length requirement. Special consideration shall be given to tees in the line being cleaned when the tee in the line is larger than the main line being cleaned. Longer stingers may be required to prevent accidental reversal of the line mole assembly in the piping.

25. All high-pressure hoses and flex lances used for line moling must be marked a minimum of 24 inches from the end of the compression fitting. This visual marking will minimize the potential for a serious water cut when the line mole on the hose end exits toward the end of the pipe.

26. All nonrotating equipment high-pressure hose assemblies must include whip checks to protect personnel from excessive movement of hoses in case of a hose rupture or blown end fitting.

27. All shotgun whip hoses must be shrouded to protect the shotgunner against accidental rupture of the whip hose. Shotgun shrouds must be constructed (e.g., double SS-braided) in a manner that protects the shotgunner against a serious water cut. The protective shroud must be a minimum of 6 feet in length. The whip hose shroud must be inspected for inner and outer integrity (e.g., excessive wear and breakdown of reinforcement wire braiding) on a quarterly basis at minimum.

28. When the system is under pressure, all personnel shall not be within 6 feet of pressurized connections unless the equipment is shielded or guarded in some manner (including weep holes).

29. The shotgun shall be a minimum of 66 inches in overall length from the butt stock of the gun to the end of the nozzle. All shotguns must be guarded against accidental activation of the gun.

30. A shotgun is required to be used to clean the first 6 inches of tube sheets and the first 24 inches of piping before a flex lance or line mole hose is utilized in the cleaning operation of exchangers or piping. This will reduce the potential for injury due to a pressurized flex lance's or line mole assembly's coming out under hydraulic pressure from the tube sheet or piping being cleaned and the high-pressure water stream coming into contact with the hydroblasting technician.

31. All equipment being cleaned shall be shielded against flying debris and chemicals that could pose a threat of potential injury or exposure to someone.

Appendix C

PROCEDURE FOR SET-UP, ADJUSTMENTS, REPAIRS, AND RIG-DOWN OF 10K AND 20K HYDROBLAST SYSTEMS OR TOOLING

A. The designated pump operator will do the following:

1. Place the pump in neutral.
2. Turn the pump engine off.
3. Turn off water to the pump at the water source supply valve.
4. Bleed residual water pressure from the hydroblast system.

B. The supervisor/crew leader in charge will do the following:

1. Verify steps A1–A4 have been completed.
2. Once those steps have been verified, direct employees to make further necessary adjustments, do repairs, rig up or down, etc.
3. Verify employees are no longer exposed to system energy and that the system is safe for restart once rig up/down, repairs, or alterations are complete.
4. Direct pump operator to restart the pump.

Procedure for Entry into Confined Space Containing Automated Tooling:

A. The designated pump operator will do the following:

1. Place the pump in neutral.
2. Turn the pump engine off.
3. Turn off water to the pump at the water source supply valve.
4. Bleed residual water pressure from the hydroblast system.

B. The supervisor/crew leader in charge will do the following:

1. Verify steps A1–A4 have been completed.
2. Apply a lock and tag to an energy isolation device at the pump that does the following:
   a. Prevents water flow to the fluid end of the pump.
   b. Prevents water flow from the fluid end by disconnecting and locking the service hose.
   c. Physically prevents electrical power from reaching the pump starter; must be a power switch that is designed to accept a lock and tag.
3. Places the key in a lockbox controlled by the confined space attendant.

C. Employees/entrants will secure tags with their name and date to the lockbox in a manner that prevents the lockbox from being opened.

D. Upon exiting the confined space, each employee must remove his/her tag from the lockbox.

E. After all the tags have been removed, the supervisor/crew leader will do the following:

1. Visually inspect the confined space.
2. Review the confined-space entry log with the attendant.
3. Verify that all employees have exited the confined space.
4. After steps E1–E3 have been completed, remove his/her key from the lockbox and return to the pump.
5. Remove his/her lock from the energy isolation device.
6. Verify with the attendant that it is safe to start the pump.
7. Direct the designated pump operator to start and pressurize the pump.

Trade Terms Introduced in This Module

Antiwithdrawal device: A clamping anti-return device for moling and flex lance operations.

Atmospheric hazard: Hazard involving gases or particulate matter affecting the air so as to create dangerous conditions.

Automation: A fully automated blasting system usually includes contained surface preparation and coating applications. Automated equipment generally consumes less water per meter than handheld equipment.

Bypass valve: A valve used as the control device to allow technicians to select the output pressure independent of the pump's actual pressure.

Capacity: Ability to contain a quantity, such as the capacity of a tank.

Cavitation: A method of coating removal. The formation, growth, and collapse of vapor-filled cavities in liquid flow.

Chemical cleaning: A surface preparation method using chemical conversion.

Chilblains: Damage to small blood vessels in the skin caused by repeated exposure to cold temperatures resulting in redness and itching.

Cold stress: Extreme exposure to cold and low-temperature conditions that leads to conditions of hypothermia, frostbite, trench foot, or chilblains.

Combustible: Capable of igniting and burning.

Confined space: Space on a work site with a size and shape that restricts the movement of anyone who must enter, work in, and exit the space.

Discharge: The side of a pump where the medium comes out.

Dump system: A valve which rapidly releases pressure from the lance.

Emergency shutdown: A procedure to stop the flow of hazardous fluid to protect people, equipment, and the work environment.

Energy isolation: Precautions to protect workers from hazardous high-energy sources, including locking and/or tagging of energy sources and disconnecting them when not in use.

Fan jet: A hydroblasting jet that spreads as it leaves the nozzle, usually at a 15-degree angle.

Fan tip: Tip of a nozzle that produces a fan-shaped flow pattern.

Filtration: Method of collecting and pumping out wastewater generated during hydroblasting.

Flame-retardant clothing: Safety gear that protects worker from a flash fire.

Flex lance: A flexible tube that attaches to the high-pressure hose, on one end, and to a stinger and/or a nozzle on the other end.

Flow rate: Quantity of liquid passing through a component in a set time; usually measured in gallons per minute (gpm), or in kilograms per square centimeter (kg/cm^2).

Flushing: Step taken before installing the tip on the nozzle to remove air bubbles and particles as well as to check the operation of the dump system. Also done to clear out any frozen water from the system during freezing conditions.

Frostbite: An injury caused by freezing that reduces blood flow to the body's extremities.

Heat cramps: A decrease in the body's salt and water level due to heat and strenuous activity that leads to muscle pain or spasms.

Heat exhaustion: A form of heat stress caused by the body losing large amounts of water and salt, usually due to excessive sweating.

Heat rash: A skin irritation caused by excessive sweating during hot, humid weather.

Heat stress: Extreme exposure to heat or work in hot environments that leads to heat-related illnesses, such as heat exhaustion or heat stroke.

Heatstroke: The most serious form of heat stress, identified by a rapid rise in body temperature, excessive sweating, an inability to cool down, hallucinations, chills, confusion, and dizziness.

Hydraulic lance: Lance that delivers a water jet generated by a hydraulic pump.

Hydroblasting technician: The person operating the lance or shotgun.

Hydroexcavation: The use of pressurized water or air and a vacuum to remove debris in a nonmechanical, nondestructive manner. It is also known as vacuum excavation, potholing, and hydrodigging.

Hypothermia: A form of cold stress when a person's body loses heat faster than it can produce it, which results in an abnormally low body temperature.

Jet: High-pressure, narrow stream of water.

Job safety analysis (JSA): Safety management assessment in which risks and hazards are identified and steps are taken to eliminate or control those hazards.

Lance: Rigid or flexible tube to which a nozzle is attached.

Lanyard: A fall-protection strap that attaches to a security point and to an individual body harness.

Line mole: Device used for clearing and cleaning pipe. A line mole has a jet spraying forward and several jets spraying back and to the side, which propels the mole down the pipe.

Maximum allowable working pressure (MAWP): The maximum pressure level that a hose is calibrated to handle.

Moling: The process of clearing a tube or pipe with a high-pressure hose and self-propelled jet nozzle.

Nonpermit-required confined space: A space that does not require a permit to enter.

Nozzle: Part of the lance that has a small aperture or opening, producing a high-speed stream of water.

Orifice: The aperture at the end of the nozzle designed to regulate flow at the end.

Oxygen-deficient atmosphere: An atmosphere with insufficient oxygen, making it difficult to breathe.

Oxygen-enriched atmosphere: An atmosphere which has a higher than usual quantity of oxygen, making the atmosphere more combustible.

Permit-required confined space: A confined space which contains some potential hazard, and which requires a permit for entry.

Plug: A tapered piece used to open or close a valve passageway.

Plunger: Displacement pump that forces fluid through a stationary high-pressure seal.

Positive displacement: Production of the same flow at a given rate at a given speed (rpm), regardless of the discharge pressure.

Pressure rating: The maximum working pressure for valves and other fittings used in high-pressure work.

Pump operator: Team member assigned to control the pump and other equipment, and to observe the hydroblasting technician and the blasting perimeter.

Quick-connect coupling: Allows tubing to be quickly connected and disconnected to reduce spills and increase safety.

Radiation exposure badge: Device worn by all workers in a hazardous area to monitor dose of radiation exposure in the area.

Rupture disk: A system or peripheral used to relieve excessive pressure in a hydroblasting setup.

Safety card: Card that describes safety issues in the hydroblasting area.

Safety shroud: Device used to guard the hydroblasting technician from failure of the hose fitting at the point where it is connected to the gun.

Scaffold: A temporary platform to support workers and materials, while providing access to work areas above the ground.

Shielding: Guarding against hose end from exiting a pipe in the event that it turns 180 degrees and is aimed back at the hydroblasting technician.

Shotgun: A lance with a rigid tube and a shoulder brace.

Spin nozzle: Nozzle used for fan jet application.

Stinger: Rigid piece of pipe affixed to a line mole to prevent reversing of the mole in the line.

Straight jet: A high-pressure stream of water that does not spread out significantly after leaving the nozzle.

Straight tip: Orifice of a nozzle that produces a straight jet of water.

Suction: Intake side of a pump.

Team leader: Person designated to manage the job, including crew set-up operations, paperwork, and communication with outsiders.

Test pressure (TP): Pressure level that is 1½ times or 20% above the standard working pressure of a particular lance.

Throttling: Decreasing flow of fluid.

Tip: Nozzle of a lance where a small aperture produces a high speed of fluid.

Trench foot: A form of cold stress caused by prolonged exposure to cold and wet conditions.

Turbulence: Chaotic, rough flow of liquid or gas, usually producing bubbles.

Ultrahigh pressure (UHP): Pressure above 15,000 psi.

Velocity: Distance traveled in a unit of time. Velocity = Function × Flow.

Water jet: High-pressure, narrow stream of water.

Weep hole: A small opening in hydroblasting connects that will not leak if properly seated.

Whip check: Restrainer attached at any point where two hoses are joined; used to control the movement of a pressurized hose if it becomes uncoupled.

Working pressure (WP): The pressure rating at which the risk level to the hydroblasting technician is acceptable.

Figure Credits

FS Solutions/Gary Toothe, Figures 2-7, 29-31, 36, 40, 44, 46, 49-53, 55-64, 66

Draeger Safety, Inc., Figure 13A

RAE Systems, Figure 13B

Topaz Publications, Figures 20, 33, 35, 37-39, 47

Ancon Marine, Figures 21, 22, 32, 34, 41, 54, 65

Gardener Denver Water Jetting System, Figure 42

NLB Corporation, Figures 43, 45

©iStockphoto.com/powerofforever, Figure 48

NCCER CURRICULA — USER UPDATE

NCCER makes every effort to keep its textbooks up-to-date and free of technical errors. We appreciate your help in this process. If you find an error, a typographical mistake, or an inaccuracy in NCCER's curricula, please fill out this form (or a photocopy), or complete the online form at **www.nccer.org/olf**. Be sure to include the exact module ID number, page number, a detailed description, and your recommended correction. Your input will be brought to the attention of the Authoring Team. Thank you for your assistance.

Instructors – If you have an idea for improving this textbook, or have found that additional materials were necessary to teach this module effectively, please let us know so that we may present your suggestions to the Authoring Team.

NCCER Product Development and Revision

13614 Progress Blvd., Alachua, FL 32615

Email: curriculum@nccer.org
Online: www.nccer.org/olf

❏ Trainee Guide ❏ AIG ❏ Exam ❏ PowerPoints Other _____

Craft / Level: _____ Copyright Date: _____

Module ID Number / Title: _____

Section Number(s): _____

Description: _____

_____

_____

_____

_____

Recommended Correction: _____

_____

_____

_____

_____

Your Name: _____

Address: _____

_____

Email: _____ Phone: _____

Index

H

Hazardous waste, 3–4. *See also* Water filtration
HAZWOPER operations, 38–39
Hearing protection, 3, 5, 6
Heat cramps, 30, 49
Heat exchangers, cleaning, 27
Heat exhaustion, 30, 49
Heat rash, 30, 49
Heat stress, 29, 49
Heatstroke, 29–30, 49
Horsepower (hp), 1
Hoses, 23–24, 31–32
Hp. *See* Horsepower (hp)
Hydraulic lance, 24, 49
Hydroblasting
   defined, 1
   operations
      bringing up the system, 36–37
      cautions, 37
      cleanup, 38
      flex lance, 37–38
      pipe cleaning, 37–38
      setup issues, 31–36
      shotgunning, 37
   process, 1
   theory underlying
      flow rate, 2
      fluid characteristics, 1–2
      impact, 2
      pressure, 2
Hydroblasting applications
   concrete repair, 26, 27–28
   cutting, 28
   hydrodemolition, 26
   hydroexcavation, 27
   industrial cleaning, 27
   pipe cleaning, 26, 37
   rigid lancing, 26
   sewer cleaning, 27
   structural cleaning, 27–28
   tank cleaning, 27, 38
   tube cleaning, 27
   typical, 1
   ultrahigh-pressure, 40
   water jetting, 26
Hydroblasting equipment
   filters, 23–24
   grounding, 28
   high-pressure, positive-displacement pump, 17–18
   hoses, 23–24
   lance, 1, 2, 3, 24–26, 50
   nozzles, 2, 26, 50
   psi ratings, standard, 1
   pumps
      classification, 19
      operating principle, 19
      reciprocating type, 19–20
      types of, 18–19
   typical, 1
   valves
      bypass valve, 22–23, 49
      common features, 20
      control methods, 20
      dry shutoff valve, 22
      dump valve, 2, 21–22
      function, 20
      materials, 20–21
      relief valves, 19, 22–23

Hydroblasting safety
   accident management, 4–5
   antiwithdrawal devices, 24–25
   back pressure, 3, 5, 28–29
   basics, 2–3, 44–47
   channeling, 4
   engine exhaust indoors, 5
   environmental issues, 3–4
   grounding equipment, 28
   infection control, 4
   noise, 3, 6
   personal protective equipment, 3, 31–32
   setup issues
      area isolation, 31–32
      beginning shift checks, 32–35
      drainage, 31
      lance hook up procedure, 35–36
      location, 31
      water supply, 32
   variances, 38–39
Hydroblasting technicians
   defined, 49
   highlighted in text, 3
   responsibilities, 28–29
   risks
      balance disruption, 5, 28
      cold stress, 30–31
      fatigue hazard, 5, 28
      heat stress, 29–30
Hydrodemolition, 26
Hydroexcavation, 27, 50
Hypothermia, 30, 50

I

Impact, water's, 2
Industrial cleaning, 27
Infection control, 4
Injuries, 4–5
Iron valves, 20

J

Jet, 2, 50
Jet stream injury, 2, 3–5
Job safety analysis (JSA), 28, 50
JSA. *See* Job safety analysis (JSA)

K

Kg/cm². *See* Kilograms per centimeter squared (kg/cm²)
Kilograms per centimeter squared (kg/cm²), 2, 43

L

Lance, 1, 24–26, 50
Lance hook up procedure, 35–36
Lance operators, 3
Lance safety, 2, 3, 24–25
Lanyard, 15, 50
Line mole, 22, 24, 27, 50
Line pulse, 22
Location setup safety, 31

M

MAWP. *See* Maximum allowable working pressure (MAWP)
Maximum allowable working pressure (MAWP), 23, 50
Medical alert cards, 4
Metal cutting, 28
Metric unit conversions to English, 43
Moling, 29, 50